Relativistic Numerical Hydrodynamics

This book presents an overview of the computational framework in which calculations of relativistic hydrodynamics have been developed. It summarizes the jargon and methods used in the field, and provides illustrative applications to real physical systems. The authors explain how to break down the complexities of Einstein's equations and fluid dynamics, stressing the viability of the Euler–Lagrange approach to astrophysical problems. The book contains techniques and algorithms enabling one to build computer simulations of relativistic fluid problems for various astrophysical systems in one, two, and three dimensions. It also shows the reader how to test relativistic hydrodynamics codes.

Suitable for use as a textbook for graduate courses on astrophysical hydrodynamics and relativistic astrophysics, this book also provides a valuable reference for researchers already working in the field.

JAMES WILSON is widely recognized as a pioneer in the field of numerical relativity and hydrodynamics. Most of the techniques currently in active use in the field today were developed by him at one stage or another. He is best known for having first solved the supernova explosion mechanism by delayed neutrino heating, as well as for developing simulations for accreting black holes, black hole and neutron star collisions, supernova jets, and binary neutron stars. In 1994 he was awarded the Marcel Grossman General Relativity Prize for his contributions to the development of the field of numerical relativity. He is also the author of numerous publications in numerical astrophysics.

GRANT MATHEWS is Professor of Theoretical Astrophysics and Cosmology at Notre Dame University, Indiana. He has been working together with Jim Wilson for the past 15 years on the development of techniques for relativistic hydrodynamics in three spatial dimensions. He has published over 200 papers in areas of theoretical and experimental astrophysics, cosmology, and relativity.

CAMBRIDGE MONOGRAPHS ON
MATHEMATICAL PHYSICS

General editors: P. V. Landshoff, D. R. Nelson, S. Weinberg

[†] Issued as a paperback

Relativistic Numerical Hydrodynamics

JAMES R. WILSON

Lawrence Livermore National Laboratory, California

GRANT J. MATHEWS

University of Notre Dame

CAMBRIDGE
UNIVERSITY PRESS

CAMBRIDGE UNIVERSITY PRESS
Cambridge, New York, Melbourne, Madrid, Cape Town, Singapore, São Paulo

Cambridge University Press
The Edinburgh Building, Cambridge CB2 8RU, UK

Published in the United States of America by Cambridge University Press, New York

www.cambridge.org
Information on this title: www.cambridge.org/9780521631556

First published 2003
This digitally printed version 2007

A catalogue record for this publication is available from the British Library

Library of Congress Cataloguing in Publication data

Wilson, James R. (James Ricker)
Relativistic numerical hydrodynamics / James R. Wilson, Grant J. Mathews.
  p.   cm.
Includes bibliographical references and index.
ISBN 0 521 63155 6
1. Relativistic fluid dynamics – Mathematical models.   2. Hydrodynamics –
Mathematical models.   I. Mathews, G. J. (Grant J.)   II. Title.
QA912.W55   2003
532′.5–dc21   2002041449

ISBN 978-0-521-63155-6 hardback
ISBN 978-0-521-03771-6 paperback

Dedicated to our loving and patient wives Demetra and Eve

Contents

Preface

We are convinced of a genuine need for a monograph describing the many facets and new developments in numerical relativistic hydrodynamics. Such calculations are crucial to several areas of current research in the physics of stellar collapse, supernovae, and black hole formation, as well as the merging of the final orbits of coalescing binary neutron stars. Both problems are only now entering the level of sophistication where three-dimensional relativistic hydrodynamics simulations are both possible and necessary. In the former problem such calculations are crucial to understand the explosion mechanism. In the latter problem, a great deal of interest in such calculations has recently been inspired by the development of next-generation gravity wave detectors to search for such events, and as a possible explanation of the physics underlying observed astrophysical γ-ray bursts.

The field of numerical relativistic hydrodynamics has developed over the past 30 years, but there has not been written a technical text explaining the many techniques relevant to this discipline, many of which are much different than standard general relativity textbook approaches. This book will present such a review of techniques for numerical general relativistic hydrodynamics developed by one of the pioneers of this field over the past three decades.

We begin by developing the equations and differencing schemes for special relativistic hydrodynamics as an introduction to the metric formulation of the problems. Here, the basic numerical techniques and a number of test problems and applications will be discussed.

Following this, the formalism for matter flows in the curved spacetime of general relativity will be presented in the usual (3+1) formalism. With the techniques established, the next chapter will then summarize cosmological applications in one spatial dimension. This will also lead naturally to

a discussion of core-collapse supernovae in spherical symmetry including the many physical complexities due to neutrino interactions and a large range of dynamic timescales.

Next we will describe some important axisymmetric problems such as stellar and black hole rotation, accretion and the head-on collision of two neutron stars. This topic also naturally leads to a discussion of magneto-hydrodynamics and its applications to axisymmetric problems.

The book then finally discusses an application in three spatial dimensions: the hydrodynamics of orbiting neutron stars. This chapter focuses on the development of the conformally flat approximation and techniques for analyzing the gravitational radiation generated by stellar collapse and binary mergers.

This book would best be described as a monograph. That is, it is a summary by experts in the field written for others at a similar level. Nevertheless, enough introductory material has been included that a graduate student or nonexpert can become familiar with the concepts without additional resource material. The main point of this book is to provide a summary of results and techniques for both the expert and nonexpert in one complete text. Some of this material has never been published and only exists in private notes. Most of the available material only exists in a number of journal publications and/or obscure conference proceedings, many of which are no longer in print.

The work described herein is of course the result of the efforts of many knowledgeable collaborators. We would particularly like to acknowledge the important contributions to the general relativistic hydrodynamics work discussed herein from Joan Centrella, Sam Dalhed, Steven Detweiler, Peter Dykema, Charles Evans, Chris Fragile, Hannu Kurki-Suonio, James LeBlanc, Pedro Marronetti, Richard Matzner, Ronald Mayle, Thomas McAbee, Jay Salmonson, and Larry Smarr. We would also like to acknowledge useful input from Peter Anninos and Dinshaw Balsara, along with help from Heidi Grantham in the preparation of some of the figures.

1
Introduction

Relativistic numerical hydrodynamics is currently a field of intense interest. On the one hand, the development of next-generation laser interferometric and cryogenic gravity wave detectors is opening a new window of astronomy, one which will peer into a world of multidimensional rapidly varying matter and gravity fields such as occur in and around neutron stars, black holes, supernovae, compact binary systems, dense clusters, collapsing stars, the early universe, etc. At the same time, X-ray and γ-ray observatories are providing (or will soon provide) a wealth of data on the evolution of matter in and around X-ray and γ-ray emitting compact objects such as accreting black holes and neutron stars. Such systems can only be realistically analyzed by a detailed numerical study of the spacetime and matter fields.

A quantitative understanding of these systems as well as a host of other astrophysical phenomena such as stellar collapse leading to supernovae, the evolution of massive stars, and the origin of γ-ray bursts, the origin and evolution of relativistic jets, all require multidimensional complex relativistic numerical simulations in three spatial dimensions. Since analytic and post-Newtonian methods are only applicable for systems of special symmetry and/or relatively weak fields, numerical relativistic hydrodynamics is the only viable method to model such highly dynamical asymmetrical strong field systems.

The technology for observing such energetic astrophysical phenomena has developed in concert with the development of high speed computing. Hence, it is perhaps no accident that the requirement for next-generation multi-dimensional relativistic hydrodynamics modeling is occurring at a time when computers are just now approaching the speed and memory capability needed to explore such systems. For these reasons, it is expected

1

that there will be much research in relativistic numerical hydrodynamics calculations in the coming years, hence the need for a book reviewing the development of the subject.

The textbooks from which most of us learn general relativity usually emphasize a number of analytic solutions of some special cases, like that of an isotropic Schwarzschild or Friedmann metric. Indeed, one is hard pressed to think of a problem in relativity which can still be addressed with paper and pencil. The remaining real-world applications in astrophysics and cosmology cannot be seriously studied analytically, nor can one ignore the hydrodynamic evolution of the matter fields. Such systems must be studied numerically. Indeed, the solution of numerical problems often requires one to abandon some or all aspects of Newtonian or even post-Newtonian intuition. Our goal here will be to provide an overview of the computational framework in which such calculations have been done, along with illustrative applications to real physical systems.

This book does not, however, attempt to give a comprehensive overview of how to do numerical relativistic hydrodynamics calculations. It is rather a compilation of those projects with which one or both of the authors have had some involvement. An attempt at a comprehensive overview of a field in which there have been so many significant contributors would be difficult. Hence, although we shall refer here to a number of other works in the field, this text will for the most part only summarize the contributions of the authors and collaborators. These are the works with which we are most familiar. Nevertheless, in the process of reading this text, it is hoped that the reader will gain some understanding of the development of relativistic hydrodynamics which has occurred over the past 30 years.

In what follows we will assume that the reader has some familiarity with basic concepts in special and general relativity. We only provide enough introductory material so that the relativistic field and matter equations can be introduced in a context which is most easily applied to numerical problems, and not in the way they might be introduced in an introductory text in either relativity or hydrodynamics alone.

1.1 Notation and convention

As with any other intensely mathematical subject, a text on numerical relativity should contain a concise summary of notation and convention in one location. Hence, we begin with an overview of the notation and conventions which we have attempted to maintain throughout the book. By and large, these are the conventions widely adopted in the field, and as such, comprise useful introductory material.

In what follows we use the usual convention of Greek indices to denote components of four-dimensional spacetime ($\mu = 0, 1, 2, 3$). When referring to a specific coordinate system they will be identified according to normal convention (e.g. $\mu = t, x, y, z$ for Cartesian coordinates). We use Latin characters, $i, j, k, \ldots$ to denote spatial indices. Partial differentiation will be denoted in both the explicit and abbreviated form, e.g.

$$\frac{\partial}{\partial x^\mu} = \partial_\mu. \tag{1.1}$$

Partial differentiation along the time coordinate will also frequently be denoted by the familiar "dot" notation,

$$\frac{\partial A}{\partial t} = \frac{\partial A}{\partial x^0} = \dot{A}. \tag{1.2}$$

We will also make use of geometrized units, $c = G = 1$. For convenience, Table 1.1 gives conversions from cgs units to geometrized units for various parameters in use in this text.

1.2 General relativity

A brief summary of general relativity is a necessary starting point for introducing concepts and notation to be encountered in subsequent chapters. General relativity derives from the principle of equivalence which asserts that at every spacetime point we can choose a coordinate system such that the laws of physics have the same form as they would in the absence of a gravitational field. This principle has led to the Einstein field equations which relate the curvature of spacetime to the distribution of mass–energy,

$$G_{\mu\nu} = 8\pi T_{\mu\nu}, \tag{1.3}$$

where $T_{\mu\nu}$ is the energy momentum (or stress energy) tensor.

The Einstein tensor $G_{\mu\nu}$ can be written in terms of the Ricci tensor $R_{\mu\nu}$, metric tensor $g_{\mu\nu}$, and Ricci scalar R,

$$G_{\mu\nu} = R_{\mu\nu} - \frac{1}{2}g_{\mu\nu}R, \tag{1.4}$$

where the Ricci tensor is a contraction of the Reimann tensor

$$R_{\mu\nu} = R^\lambda{}_{\mu\lambda\nu}, \tag{1.5}$$

and

$$R = g^{\mu\nu}R_{\mu\nu}. \tag{1.6}$$

Table 1.1 Conversion from cgs units to geometrized units

Quantity	cgs units	Geometrized units
Length	l (cm)	(cm)
Time	t (s)	$ct = 2.99792458 \times 10^{10} \times t$ (cm)
Velocity	V (cm s^{-1})	$v/c = v/2.9979245 \times 10^{10}$ (dimensionless)
Mass	m (g)	$Gm/c^2 = 7.42426 \times 10^{-29} \times m$ (cm)
Energy	e (erg)	$Ge/c^4 = 8.26060 \times 10^{-50} \times e$ (cm)
Internal energy	ϵ (erg g^{-1})	$\epsilon/c^2 = 1.11265 \times 10^{-21} \times \epsilon$ (dimensionless)
Mass density	ρ (g cm^{-3})	$G\rho/c^2 = 7.42426 \times 10^{-29} \times \rho$ (cm^{-2})
Energy density	$\rho\epsilon$ (erg cm^{-3})	$G\rho\epsilon/c^4 = 8.26060 \times 10^{-50} \times \rho\epsilon$ (cm^{-2})
Pressure	P (dyn cm^{-2})	$GP/c^4 = 8.26060 \times 10^{-50} \times P$ (cm^{-2})
Temperature	T (K)	$GkT/c^4 = 8.26060 \times 10^{-50} \times kT$ (cm)
Entropy	S/k (dimensionless)	S/k (dimensionless)
Angular momentum	J (g cm^2 s^{-1})	$GJ/c^3 = 2.47647 \times 10^{-39} J$ (cm^2)
Angular frequency	ω (rad s^{-1})	$\omega/c = 3.33564 \times 10^{-11}\omega$ (cm^{-1})
Luminosity	L (erg s^{-1})	$GL/c^5 = 2.75544 \times 10^{-60} L$ (dimensionless)
Solar mass	$M_\odot = 1.989 \times 10^{33}$ g	$GM_\odot/c^2 = 1.477$ km
Solar luminosity	$L_\odot = 3.826 \times 10^{33}$ erg s^{-1}	$GL_\odot/c^5 = 1.054 \times 10^{-26}$ (dimensionless)
Nuclear density	$\rho_N = 2.67 \times 10^{14}$ g cm^{-3}	$G\rho_N/c^2 = 1.98 \times 10^{-14}$ (cm^{-2})

Here, the Riemann–Christoffel curvature tensor $R^\sigma{}_{\mu\kappa\nu}$ is

$$R^\sigma{}_{\mu\kappa\nu} = \partial_\kappa \Gamma^\sigma{}_{\mu\nu} - \partial_\nu \Gamma^\sigma{}_{\mu\kappa} + \Gamma^\sigma{}_{\eta\kappa}\Gamma^\eta{}_{\mu\nu} - \Gamma^\sigma{}_{\eta\nu}\Gamma^\eta{}_{\mu\kappa}, \qquad (1.7)$$

where the Christoffel symbols, $\Gamma^\alpha{}_{\mu\nu}$, relate directly to the metric tensor. For the usual case of a coordinate system in which the basis vectors commute we have,

$$\Gamma^\alpha{}_{\mu\nu} = \frac{1}{2} g^{\alpha\beta} \left\{ \frac{\partial g_{\beta\mu}}{\partial x^\nu} + \frac{\partial g_{\beta\nu}}{\partial x^\mu} - \frac{\partial g_{\mu\nu}}{\partial x^\beta} \right\}. \qquad (1.8)$$

1.2.1 Metric tensor

From Eqs. (1.4)–(1.8) we see that the geometry of spacetime is specified once the metric $g_{\mu\nu}$ and its derivatives are given. The generalization from special relativity to general relativity is then simply the generalization from a Euclidean flat space metric tensor to a curved space metric. As in special relativity, the infinitesimal proper interval between two events in spacetime is denoted

$$ds^2 = g_{\mu\nu} dx^\mu dx^\nu. \qquad (1.9)$$

Now however, $g_{\mu\nu}$ can no longer be described by a simple Minkowski metric, but instead involves curvature. In this book we will consistently use the Misner, Thorne and Wheeler [13] sign conventions whereby the Einstein equation and the Riemann tensor have a positive sign as written above and $g_{\mu\nu}$ is *space like*, e.g. in flat space,

$$g^{flat}_{\mu\nu} = \eta_{\mu\nu} = \begin{pmatrix} -1 & 0 & 0 & 0 \\ 0 & 1 & 0 & 0 \\ 0 & 0 & 1 & 0 \\ 0 & 0 & 0 & 1 \end{pmatrix}. \qquad (1.10)$$

1.2.2 Energy momentum tensor

The other side of the Einstein equation (1.3), and the part of most interest for relativistic hydrodynamics, is of course the stress energy tensor, $T_{\mu\nu}$. In a frame of reference in which a perfect fluid is in motion with respect to an observer, the energy momentum tensor is written most generally as

$$T_{\mu\nu} = (\rho + \rho\epsilon + P)U_\mu U_\nu + P g_{\mu\nu}, \qquad (1.11)$$

where ρ is the local baryon rest-mass energy density. ρ is related to the baryon number density n_b

$$\rho = m_0 n_b, \qquad (1.12)$$

where m_0 is the baryon rest mass appropriate to the matter composition. The quantity ϵ contains all information about the net internal energy per unit mass of the baryons. It can be less than zero, for example for a nondegenerate gas of bound nuclei. The quantity P in Eq. (1.11) is the pressure, and U_μ is the covariant four-velocity. In a reference frame that is at rest and locally Lorentzian, the stress energy tensor for an isotropic perfect fluid can be written in a familiar form,

$$T_{\mu\nu} = \begin{pmatrix} \rho(1+\epsilon) & 0 & 0 & 0 \\ 0 & P & 0 & 0 \\ 0 & 0 & P & 0 \\ 0 & 0 & 0 & P \end{pmatrix}. \tag{1.13}$$

However, this form is not particularly useful for hydrodynamic simulations with nontrivial fluid motion. In what follows we will deal almost exclusively with Eq. (1.11), correcting for imperfect fluids and other fields (e.g. electromagnetic) where appropriate.

1.2.3 Covariant differentiation

Equations of motion in general relativity require the introduction of covariant differentiation. We use the notation $A^\mu{}_{;\nu}$ or $D_\nu A^\mu$ to denote covariant differentiation of a contravariant vector A^μ,

$$D_\nu A^\mu = A^\mu{}_{;\nu} = \frac{\partial A^\mu}{\partial x^\nu} + \Gamma^\mu{}_{\nu\lambda} A^\lambda. \tag{1.14}$$

Similarly, the covariant derivative of a covariant vector is

$$A_{\mu;\nu} = \frac{\partial A_\mu}{\partial x^\nu} - \Gamma^\lambda{}_{\mu\nu} A_\lambda. \tag{1.15}$$

In what follows we will also introduce covariant differentiation in the ADM three-space (cf. Section 1.3). This we denote:

$$D_i A^j = \frac{\partial A_i}{\partial x^j} + \Gamma^i{}_{jk} A^k, \tag{1.16}$$

where $\Gamma^i{}_{jk}$ now denotes connection coefficients for the three-dimensional ADM hypersurface [22].

$$\Gamma^i{}_{jk} = \frac{1}{2}\gamma^{il}\left\{\frac{\partial\gamma_{lj}}{\partial x^k} + \frac{\partial\gamma_{lk}}{\partial x^j} - \frac{\partial\gamma_{jk}}{\partial x^l}\right\}$$
$$+ \frac{1}{2}\left(\gamma^{il}\gamma_{km}C^m{}_{lj} + \gamma^{il}\gamma_{jm}C^m{}_{lk} - C^i{}_{jk}\right), \tag{1.17}$$

where the $C^i{}_{jk}$ enter when the basis vectors of the three-space (denoted e_i) do not commute, $[e_i, e_j] = e_k C^k{}_{ij} \neq 0$. For most applications, the simplifications embodied in commuting coordinates are desirable so that we can take $C^k{}_{ij} = 0$ and the three-space connection coefficient simply becomes the usual Christoffel symbol defined in terms of γ_{ij}.

The covariant derivative of a scalar α is just the ordinary gradient

$$\alpha_{;\mu} = \frac{\partial \alpha}{\partial x^\mu}. \tag{1.18}$$

The generalization of covariant differentiation to tensors of higher rank is straightforward. For each contravariant index μ a term with $\Gamma^\mu{}_{\nu\lambda}$ times the tensor is added, but with μ in the tensor replaced by λ. For each covariant index ν one subtracts a term with $\Gamma^\kappa{}_{\nu\lambda}$ times the tensor with ν replaced with κ. For example,

$$D_\rho T^\mu{}_\nu = T^\mu{}_{\nu;\rho} = \frac{\partial T^\mu{}_\nu}{\partial x^\rho} + \Gamma^\mu{}_{\rho\lambda} T^\lambda{}_\nu - \Gamma^\kappa{}_{\nu\rho} T^\mu{}_\kappa. \tag{1.19}$$

A particularly useful operation when deriving the hydrodynamic equations of motion is the covariant divergence. For a vector this simplifies to

$$V^\mu{}_{;\mu} = \frac{1}{\sqrt{g}} \frac{\partial}{\partial x^\mu} \left(\sqrt{g} V^\mu \right). \tag{1.20}$$

For a tensor it simplifies to

$$T^{\mu\nu}{}_{;\mu} = \frac{1}{\sqrt{g}} \frac{\partial}{\partial x^\mu} \left(\sqrt{g} T^{\mu\nu} \right) + \Gamma^\lambda{}_{\mu\nu} T^{\mu\nu}, \tag{1.21}$$

where g is the negative of the determinant of the metric tensor

$$g \equiv -\det(g_{\mu\nu}). \tag{1.22}$$

The above relations extend trivially to covariant differentiation in the ADM three-space by simply writing them in terms of spatial indices. In what follows, we will usually denote the determinant of the three-metric by

$$\gamma^2 \equiv \det(\gamma_{ij}). \tag{1.23}$$

1.2.4 Bianchi identities

The Riemann curvature tensor obeys some special symmetries known as the Bianchi identities,

$$R_{\lambda\mu\nu\kappa;\eta} + R_{\lambda\mu\eta\nu;\kappa} + R_{\lambda\mu\kappa\eta;\nu} = 0. \tag{1.24}$$

The contracted form of this equation implies [19] that the covariant divergence of the Einstein tensor also vanishes

$$G^{\mu\nu}{}_{;\nu} = 0. \tag{1.25}$$

This, together with the Einstein equation (1.3), immediately implies,

$$T^{\mu\nu}{}_{;\nu} = 0, \tag{1.26}$$

from which the fluid equations of motion will be derived in subsequent chapters.

In principle this is all we need to describe any system. The problem, however, is that the Einstein equations as they stand are too general. What is needed is a reduction of the Einstein equations to a form suitable for evolving in time. This we now describe.

1.3 (3+1) or ADM formalism

In general relativity and special relativity the distinction between spatial three dimensions and the time dimension becomes obscure. Indeed, it is an intriguing and beautiful aspect of relativity that what one means by time and space depends upon which observer is making the determination. However, in spite of this ambiguity, one wishes to have a method to describe a system which is "evolving in time," as this is the way we perceive events in Nature.

A way to approach this problem which nearly aligns with Newtonian intuition is to construct successive "snapshots" or "time slices" of the spatial (curved) three-space geometry along a sequence of steps in a time-like coordinate t. That is, spacetime is sliced (or foliated) into a one-parameter family of hypersurfaces separated by differential displacements in coordinate time t.

Indeed, there are many ways in which one could follow a time-like dimension and watch events unfold in spacetime. However, the one which most nearly aligns with Newtonian intuition (and one which is usually amenable to numerical methods) is the ADM (after its inventors Arnowitt, Deser and Misner [3]) or $(3 + 1)$ formalism [13].

In this approach, the time evolution of the metric is expressed as first-order time derivatives, while $G_{\mu\nu}$ contains second-order time derivatives. The time-like coordinate is chosen along a normal to the space-like hypersurfaces. Figure 1.1 shows a two-dimensional spacetime depiction of one way to do this division. That is, space and time are placed on separate footings by first specifying the proper time interval $d\tau$ between the lower and upper hypersurfaces along the direction of the normal $\vec{n}$ to

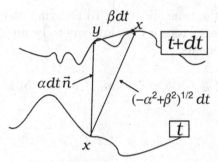

Fig. 1.1. Schematic depiction of the ADM metric.

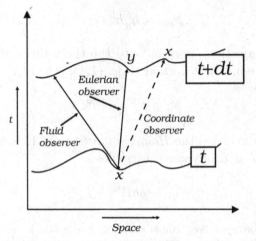

Fig. 1.2. Schematic depiction of observer frames in the ADM metric.

the spatial hypersurfaces,

$$\vec{d\tau} = dt\vec{n}. \tag{1.27}$$

We call an observer in this frame, the *Eulerian observer* as depicted in Figure 1.2.

1.3.1 Eulerian observer

An Eulerian observer moves through spacetime in a direction orthogonal to all spacetime vectors confined to the spatial hypersurfaces. This frame does not necessarily move along a force-free trajectory or geodesic in spacetime. Since the Eulerian frame is defined independently of the coordinates chosen, it is a good frame in which to measure physical quantities such as fluid velocity, neutrino energy, etc. The Einstein equations are solved in this Eulerian frame by projecting them onto the

the unit normal to the time slices n_μ. To Eulerian observers, the perfect-fluid ADM stress energy tensor then appears to be an imperfect fluid with the following components:

$$T_{\mu\nu} = dn_\mu n_\nu + s_\mu n_\nu + n_\mu s_\nu + S_{\mu\nu}, \tag{1.28}$$

where

$$\rho_H \equiv n^\mu n^\nu T_{\mu\nu}, \tag{1.29}$$

$$s_\mu \equiv n^\delta h_\mu^\beta T_{\beta\delta}, \tag{1.30}$$

$$S_{\mu\nu} = h_\mu^\delta h_\nu^\beta T_{\delta\beta}. \tag{1.31}$$

Here, h_μ^ν is a projection operator onto the three-slices and is written in terms of the time-like unit normal (Eqs. (1.44), (1.45)),

$$h_\mu^\nu \equiv \delta_\mu^\nu + n_\mu n^\nu. \tag{1.32}$$

The quantity ρ_H is called the *Hamiltonian density*. It is an ADM matter density related to hydrodynamic variables as

$$\rho_H = \rho h W^2 - P, \tag{1.33}$$

where ρ is the baryon rest mass energy, Eq. (1.12). The quantity h is called the *specific enthalpy*. Written in terms of the internal energy per gram of material ϵ and the pressure P it is

$$h \equiv 1 + \epsilon + P/\rho. \tag{1.34}$$

The quantity $W \equiv \alpha U^t$ is a generalized Lorentz factor described below.

The quantity s_μ is the *ADM momentum density*. In covariant form its spatial components are equivalent to the spatial components of the relativistic four-momentum density $S_\mu \equiv \rho_h W U_\mu$,

$$s_i = S_i = \rho h W U_i. \tag{1.35}$$

The space–space component of $S_{\mu\nu}$ is called the *spatial stress*. For a perfect-fluid stress energy tensor, the spatial stress reduces to

$$S_{ij} = P\gamma_{ij} + \rho h U_i U_j = P\gamma_{ij} + \frac{S_i S_j}{\rho h W^2}. \tag{1.36}$$

1.3.2 Coordinate observer

Next, one must also have a description of proper spatial distance dl^2 between one point and another within the three-space, i.e.

$$dl^2 = \gamma_{ij}dx^i dx^j, \tag{1.37}$$

where γ_{ij} is now the metric of the spatial three-geometry. An observer at rest in the coordinates of the spatial hypersurfaces we will refer to as the *coordinate observer* as depicted in Figure 1.2. The coordinate observer moves through the spacetime in a direction such that the spatial coordinates x^i on slice t are equal to the space coordinates of his location at time $t+dt$. In the ADM formalism, one has a freedom to choose where those coordinates connect to the next spatial hypersurface. That is, there is a shift $\vec{\beta}$ of spatial coordinates from one three-space to the next. Denoting the coordinate as y^i the point where the normal to the hypersurface (Eulerian observer) connects with the next slice (cf. Figure 1.2), we have

$$x^i(t) = x^i(t+dt) = y^i + \beta^i dt, \tag{1.38}$$

as noted on Figure 1.1. The length of the spacetime vector stretching from the coordinate observer's position at time t to that at time $t+dt$ is then that denoted by the observer's watch,

$$ds_{CO}^2 = (-\alpha^2 + \beta_i\beta^i)dt^2, \tag{1.39}$$

so that the lapse of proper time on a clock moving in the coordinate observer's frame is then given by

$$\tau = \sqrt{\alpha^2 - \beta^2}dt. \tag{1.40}$$

1.3.3 ADM metric

With these three ingredients, the proper spacetime interval between any two points in the four-space can be written,

$$ds^2 = -(\alpha^2 - \beta_i\beta^i)dt^2 + 2\beta_i dx^i dt + \gamma_{ij}dx^i dx^j. \tag{1.41}$$

The ADM metric tensor can then be written

$$g_{\mu\nu} = \begin{pmatrix} (\beta_l\beta^l - \alpha^2) & \beta_j \\ \beta_i & \gamma_{ij} \end{pmatrix}. \tag{1.42}$$

The contravariant form for the metric then becomes,

$$g^{\mu\nu} = \begin{pmatrix} -\frac{1}{\alpha^2} & \frac{\beta^i}{\alpha^2} \\ \frac{\beta^j}{\alpha^2} & \gamma^{ij} - \frac{\beta^i\beta^j}{\alpha^2} \end{pmatrix}. \tag{1.43}$$

The time-like normal vector to the three-space can thus be written,

$$n_\mu = (-\alpha, 0, 0, 0),$$ (1.44)

$$n^\mu = \left(\frac{1}{\alpha}, -\frac{\beta^i}{\alpha}\right),$$ (1.45)

where we have

$$n^\mu n_\mu = -1.$$ (1.46)

With this splitting of the metric we can think of the three-space represented by γ_{ij} as surfaces or slices which evolve in coordinate "time" as depicted in Figure 1.1.

1.3.4 Fluid observer

Before leaving this discussion on the ADM metric and observer frames it is useful to define one more frame, that of the *fluid observer*. The fluid observer (or *Lagrangian observer*) rides along with the fluid. This is the frame in which to measure the fluid rest-mass density and temperature, etc. without the complications induced by moving coordinates (e.g. different temperatures in different directions). The fluid four-velocity will be denoted as U^μ and is defined relative to the Eulerian frame. The four-velocity satisfies the condition,

$$U^\mu U_\mu = -1.$$ (1.47)

In this formalism we define the matter spatial three-velocity

$$V^\mu \equiv \frac{U^\mu}{U^t} = (1, V^i),$$ (1.48)

and introduce a Lorentz-like scalar,

$$W \equiv -n_\mu U^\mu = \alpha U^t.$$ (1.49)

It follows from $U_\mu U^\mu = -1$ that,

$$W^2 = 1 + \gamma^{ij} U_i U_j = 1 + \gamma_{ij} U^i U^j = \frac{1}{1 - \gamma_{ij} V^i V^j / \alpha^2}.$$ (1.50)

The quantity W is then the general relativistic analog of the special relativistic "gamma" factor that the Eulerian observer would assign to matter moving in the fluid frame. The vector V^i is the curved spacetime generalization of the flat space special relativistic three-velocity.

1.3.5 Field equations in (3+1) formalism

In writing the ADM metric we are free to impose some functional constraints on α and $\vec{\beta}$ and/or γ_{ij} as described below. Their choice represents a choice of coordinates referred to as a *gauge choice*. We apply specific gauge choices and derive the resultant ADM equations in subsequent chapters. Here we give the field equations in general form and summarize some of the common gauge choices.

The application of the ADM metric (1.41) to the Einstein equation (1.3) leads to a set of dynamic and constraint equations for the metric components. As we shall see, this is somewhat analogous to Maxwell's equations in classical electrodynamics.

Dynamic equations. The dynamic equations come from the $G^{t\mu}$ part of the Einstein equation (1.3). The evolution of the three-metric γ_{ij} is given by

$$\dot{\gamma}_{ij} = -2\alpha K_{ij} + D_i\beta_j + D_j\beta_i, \tag{1.51}$$

where K_{ij} is the conjugate variable to the three-metric. It is called the *extrinsic curvature*. Recall that our slices involve infinitesimal coordinate transformations along the normal vector n^μ. In general, the effect of an infinitesimal coordinate transformation on any tensor is that the new tensor equals the old tensor at the same coordinate point plus the Lie derivative of the tensor. Thus, K_{ij} is defined as the Lie derivative of γ_{ij} along the vector n^μ. The effect of a Lie derivative on a tensor $T_{\mu\nu}$ along a vector field v^λ is,

$$\mathcal{L}_v = v^\lambda T_{\mu\nu\ ;\lambda} + T^\lambda{}_\mu v_{\lambda\ ;\nu} + T^\lambda{}_\nu v_{\lambda\ ;\mu}. \tag{1.52}$$

Thus, K_{ij} is defined from the Lie derivative of γ_{ij} along the vector n^μ. It represents a motion of the three-metric,

$$K_{ij} = -\frac{1}{2}\mathcal{L}_n\gamma_{ij}, \tag{1.53}$$

where,

$$\mathcal{L}_n\gamma_{ij} = \left(\frac{1}{\alpha}\frac{\partial}{\partial t} + \mathcal{L}_\beta\right)\gamma_{ij}. \tag{1.54}$$

This is how the β terms enter into Eq. (1.51). The extrinsic curvature can also be thought of as describing the distortion of a curve when passing from one coordinate time slice to the next [13, 22].The derivation of the dynamic equation for K_{ij} is straightforward but tedious [8, 13, 22]. The result is,

$$\dot{K}_{ij} = - D_i D_j \alpha + \alpha \left[R_{ij} - 2K_{il}K^l{}_j + KK_{ij} - 8\pi S_{ij} - 4\pi\gamma_{ij}\left(\rho_H - S\right)\right]$$
$$+ \beta^l D_l K_{ij} + K_{il}D_j\beta^l + K_{lj}D_i\beta^l, \tag{1.55}$$

where R_{ij} is the Ricci tensor (1.5).

The quantity S in Eq. (1.55) is the trace of the spatial stress,

$$S \equiv \mathrm{tr}(S_{ij}) = S^i{}_i = 3P + \rho h(W^2 - 1). \tag{1.56}$$

Similarly, K is the trace of the extrinsic curvature,

$$K \equiv \mathrm{tr}(K_{ij}) = K^i{}_i. \tag{1.57}$$

From the trace of Eq. (1.55) we have the following dynamic equation for K,

$$\dot{K} = -D_i D^i \alpha + \alpha\left[R + K^2\right] + 4\pi\alpha[S - 3\rho_H] + \beta^l D_l K. \tag{1.58}$$

The Einstein equations are second order in time and space derivatives. The (3+1) decomposition above has reduced the time evolution equations to just two first-order equations in time. Nevertheless, their solution is exceedingly difficult and there exist inherent nonlinearities in the dynamic equations [1, 2] that have slowed application of this formalism in three spatial dimensions (although there has been considerable recent progress [7]). Recently proposed reformulations of the problem in terms of auxiliary variables and/or hyperbolic equations show some promise [4, 14] (see Chapter 7). A discussion of this method is beyond the scope of the present text, however, and will not be considered further here.

1.3.6 Constraint equations

The constraint equations come from projecting the Einstein equation (1.3) along n^μ [23]. This results in the *Hamiltonian constraint*

$$G_{\mu\nu}n^\mu n^\nu = \frac{1}{2}\left(R + K^2 - K_{ij}K^{ij}\right) = 8\pi\rho_H, \tag{1.59}$$

and the *momentum constraint*

$$-(\delta^i_\mu + n^i n_\mu)G^\mu{}_\nu n^\nu = D_j(K^{ij} - \gamma^{ij}K) = 8\pi s^i. \tag{1.60}$$

where s^i is the ADM contravariant three-momentum density (Eq. (1.30)). It is important to keep the distinction in mind [9, 11] between this quantity and the spatial components of the four-momentum which we will solve for in the hydrodynamic equations. In contravariant form the two quantities are not the same.

We can now see the analogy between the ADM formulation of gravity and Maxwell's electromagnetism. Both the ADM and Maxwell's equations can be written as two constraint equations plus two dynamic equations. In electromagnetism the constraint equations for electric and magnetic fields are embodied in the $\nabla \cdot E$ and $\nabla \cdot B$ equations, while the dynamic equations are contained in Ampère's law and Faraday's law. In relativity the analogous constraint equations are the ADM momentum and Hamiltonian constraints. The dynamic equations are the ADM $\dot{K}_{ij}$ and $\dot{\gamma}_{ij}$ equations. In either electromagnetism or gravity, any field configuration which satisfies the constraint equations alone represents a valid solution. The solution of the ADM equations consists of first finding such a solution to the constraint equations (known as the *initial value problem*) and then evolving the dynamic equations. A numerical solution to the dynamic equations should preserve the constraints.

1.3.7 Solving the ADM equations

Note that we have ten functions which must be solved to describe the metric. They are α, the three β^i, and the six independent components of γ_{ij}. To derive these we have the ten ADM equations. These are the Hamiltonian constraint plus the three momentum constraint equations as well as the six independent components of the $\dot{\gamma}_{ij}$ and/or $\dot{K}_{ij}$ equations.

We have a freedom to choose coordinates which is a freedom to impose four functional relations amongst the metric functions. In the (3+1) formalism this is usually done by choosing the four functions α and β^i by some prescription, although one can also specify restrictions on the three-metric γ_{ij}. In whatever form these four relations are chosen, they are the *gauge choices*. Some examples of various gauges are given below and in subsequent chapters. As we shall see, the art of solving relativistic hydrodynamics often involves the art of choosing the best gauge.

Given the four functional relations of the gauge choice, α and β^i are solved along with the evolution of the $\dot{\gamma}_{ij}$ or $\dot{K}_{ij}$ equations subject to the four constraint equations. Two degrees of freedom are then left. They can be thought of as representing the two polarizations of a gravity wave.

There is, however, no simple way to avoid the problem of evolving γ_{ij} and/or K_{ij} numerically, while also satisfying the constraint equations.

Three different approaches to that problem in particular have been proposed. They are as follows.

Solution I – initial value plus evolution. In this method one begins by finding an initial metric such that the constraint equations are satisfied (at least at the numerical level). Then the $\dot{\gamma}_{ij}$ and/or $\dot{K}_{ij}$ equations (1.51), (1.55) are explicitly evolved in time. As the calculation proceeds one can check to see how well the constraint equations remain satisfied by evaluating the Hamiltonian error,

$$\mathcal{E}_H = \frac{\int |2\rho_H - R - K^2 - K_{ij}K^{ij}|dV}{\int 2\rho_H dV}, \tag{1.61}$$

and the momentum constraint error,

$$\mathcal{E}_M = \frac{\int \left(S^i - D_j(K^{ij} - \gamma^{ij}K) \right)\left((S_i - D^j(K_{ij} - \gamma_{ij}K) \right)dV}{\int S^i S_i dV}. \tag{1.62}$$

When $\mathcal{E}_H$ and/or $\mathcal{E}_M$ become too large (say a few percent), give up. This is analogous to the hydrodynamics calculations described in subsequent chapters where the total energy is evaluated at various times. When total energy nonconservation becomes too large one no longer has confidence in the results.

Solution II – mixed scheme. In this scheme constraint equations are solved at each time step for a subset of the functions γ_{ij} and/or K_{ij}. The complementary set is evolved by the appropriate $\dot{\gamma}_{ij}$ and/or $\dot{K}_{ij}$ equations. If one could separate out the gravity wave degrees of freedom, this would be an ideal solution. However, this approach has not yet been successfully implemented except for axisymmetric systems.

Solution III – constrained metric condition. In this approach, the gravity waves are ignored and an additional condition is imposed upon the metric such that all metric components are defined. This is the essence of the conformally flat condition discussed in Chapter 7. Although repulsive to the purist, this method has been shown to provide surprisingly accurate results and is much easier to implement in three spatial dimensions than the previous two schemes.

1.3.8 Gauge choices for γ_{ij}

In view of the difficulties in applying the ADM method, a large fraction of the work has involved special restrictions on the three-metric. Some common gauge choices based upon various symmetries are summarized as follows.

1. Spherical symmetry

This is the best known simplification. In this system Birkhof's theorem applies. That is, it can be shown that any spherically symmetric vacuum solution of Einstein's equations must be static and must agree with the Schwarzschild solution. Nevertheless, other coordinate choices (which transform to Schwarzschild coordinates) can and have been utilized to particular advantage. It is of course best to choose a coordinate system that matches the problem one wishes to solve. Spherical coordinates r, θ, ϕ are the obvious choice for problems with spherical symmetry. The most general three-metric in spherical spacetime can be written as

$$dl^2 = A dr^2 + B r^2 (d\theta^2 + \sin^2 \theta d\phi^2), \tag{1.63}$$

where A and B are functions of t and r only. For example, in standard static Schwarzschild coordinates

$$A = (1 - 2M/r)^{-1}, \tag{1.64}$$

and $B = 1$. In isotropic coordinates

$$A = B = \left(1 + \frac{M}{2r}\right)^4, \tag{1.65}$$

where M is the mass interior to r.

In dynamical calculations of stellar collapse, it is useful to utilize mass instead of r as the spatial coordinate. This is usually accomplished via an adaptation of the *May and White metric* [12], whereby the spatial metric can be written,

$$dl^2 = B^2 dm^2 + r^2 (d\theta^2 + \sin^2 \theta d\phi^2). \tag{1.66}$$

An application of this metric to stellar collapse is discussed in Chapter 5. The other metric variables are at most α and β^r. In spherical geometry these also are only functions of t and r. Their choice is discussed below.

2. Planar symmetry

This is the simplest geometry with nontrivial gravity waves. Its application in cosmology is discussed in Chapter 4. The field equations in these coordinates are simple enough that the gravity wave aspect is solvable. Constraints are less tightly controlled by gravity waves. Since more than one dimension is necessary to describe gravity waves, the metric choice in planar cosmologies, for example, can be written [6]

$$dl^2 = A^2 \left[dx^2 + h^2 dy^2 + dz^2 \right], \tag{1.67}$$

where the coefficients A and h are coordinates of t and z only. Equations for the shift vector components β^x and β^z are then derived to maintain these coordinates.

3. Axial symmetry

This problem has been extensively worked on. The mixed scheme has been used with sufficiently good numerical techniques that satisfactory results have been obtained even with modest computer resources for the evolution of $\dot{\gamma}_{ij}$ and $\dot{K}_{ij}$ [8, 17][8, 17, 20].

In axial symmetry r, z, ϕ or r, θ, ϕ coordinates have been most often used, depending upon the problem one wishes to solve, e.g. head-on neutron star collisions or axially symmetric stellar collapse. In some cases bipolar or bispherical coordinates may be applicable.

In stationary spacetime, the simplest case is that of a static, nonrotating, vacuum axisymmetric line element. In this case the Weyl three-metric can be chosen for which $\gamma_{rr} = \gamma_{zz}$, and the three-line element becomes

$$dl^2 = A^2(dr^2 + dz^2) + r^2 B^2 d\phi^2. \tag{1.68}$$

In a time-dependent nonrotating axisymmetric spacetime, the most general three-metric for r, z, ϕ coordinates can be written

$$dl^2 = Adr^2 + Bdz^2 + Cdrdz + DR^2 d\phi^2, \tag{1.69}$$

where $\gamma_{r\phi} = \gamma_{z\phi} = 0$, and $\beta^\phi = 0$.

4. No symmetry

In problems with no angular symmetry, simple Cartesian x, y, z coordinates are the most straightforward choice, and in general all six components of the three-metric γ_{ij} must be evolved. As a solution to the initial value problem, however, one often begins by the implementation of a *conformally flat condition*. The three-metric is then constrained such that the spatial line element becomes

$$dl^2 = \phi^4[dx^2 + dy^2 + dz^2]. \tag{1.70}$$

The six components of the three-metric are thus reduced to a single position-dependent *conformal factor* ϕ. As will be discussed in Chapter 7, good numerical results can be obtained in this way even for some dynamical systems. While this choice is allowable along any individual time slice, it does not allow for gravity wave propagation and does not reproduce a Kerr solution. A modified metric which does reduce to the correct Kerr limit has been applied recently by Usui, Uyru and Eriguchi [18]. For this case the metric is chosen to be:

$$ds^2 = -e^{2\nu}dt^2 + r^2\sin^2\theta e^{2\beta}(d\phi - \omega dt)^2 + e^{2\alpha}dr^2 + r^2 e^{2\alpha'}d\theta^2. \tag{1.71}$$

This is a simplification because the off-diagonal $t - r$ and $t - \theta$ metric coefficients are neglected. Only the dominant $t - \phi$ term is kept.

Gauge choice for α. As mentioned above it is most common also to apply gauge conditions on the metric coefficients α and/or β^i. The simplest possible gauge condition is to use time-like geodesics normal to the space-time slices and to label the slices with the local proper time of the Eulerian observer; that is, simply to set $\alpha = 1$, $\beta^i = 0$. This is the *synchronous gauge*. This choice is fine for mildly relativistic systems. However, these frames tend to develop coordinate singularities by focusing of the normal geodesics [22]. This choice is also difficult to relate to Newtonian physics in the strong field regime.

A much more frequently invoked choice is what is referred to as the *maximal slicing condition* [10, 22]. That is, one simply sets,

$$\text{tr}(K_{ij}) \equiv K^i{}_i \equiv K = 0. \qquad (1.72)$$

This gauge acts as an effective anti-focusing condition which avoids the coordinate singularities of the synchronous gauge. Its name derives from the fact that the volume $V(S)$ of any bounded portion S of an ADM slice,

$$V(S) = \int_S (\det \gamma_{ij})^{1/2} d^3x, \qquad (1.73)$$

can be shown [22] to be a maximum. One can then take the trace of the $\dot{K}_{ij}$ equation (1.55) to find an equation for α.

For strong fields in this gauge, i.e. when $\rho_\alpha r^3 \gg 1$, it can be shown that $\alpha \to 0$ exponentially. Since the right-hand side of the $\dot{\gamma}_{ij}$ (Eq. (1.51)) and $\dot{K}_{ij}$ (Eq. (1.55)) equations is multiplied by α this implies that the gravity evolution ceases around regions of high mass–energy density. Thus black holes can form and the evolution can continue outside the black holes.

The maximal slicing condition is appropriate for asymptotically flat spacetimes. However, in closed universes there are compact slices without a boundary. Hence, at most only one maximal slice exists. In this case, an equally simple choice is that of constant in space mean extrinsic curvature, referred to as *closed slices*, i.e.

$$\text{tr}(K_{ij}) = \text{constant}, \qquad (1.74)$$

where a different positive constant can apply for each slice. One is also free to fix $\partial_t \text{tr}(K_{ij}) = \text{constant}$ for each slice. If $\partial_t \text{tr}(K_{ij}) = 1$ for all slices, then one can identify the trace with coordinate time, $\text{tr}(K_{ij}) = t^{-1}$. Then K becomes as a natural time coordinate. This is a convenient choice in cosmology.

Another possibility [22] is for $\mathrm{tr}(K_{ij})$ to equal a negative constant. This is possible in some asymptotically flat spacetimes (e.g. Schwarzschild). Such slices are similar to the *mass hyperboloids* of Minkowski spacetime. This is referred to as *asymptotically null slicing*. In this gauge, the lapse α grows like r at large distances.

1.3.9 Gauge choices for $\vec{\beta}$

Having utilized a slicing condition to obtain α one must still choose a shift vector $\vec{\beta}$. That is, one must still select how the coordinates are moved from one coordinate time slice to the next. The simplest choice is of course $\vec{\beta} = 0$, which has been dubbed the *Eulerian gauge* [15, 16] or the *normal gauge* because the spatial coordinates are constant along the normal or Eulerian trajectories. This choice simplifies the equations, however, it has been learned that a better choice is to select $\vec{\beta}$ in such a way as to simplify the the numerics and/or the interpretation of the results.

In hydrodynamics, it is often useful to allow the spatial coordinate frames to move with the matter. This is referred to as the *comoving gauge* or the *Lagrangian gauge*, for which one writes,

$$\beta^i = \alpha V^i, \tag{1.75}$$

where V^i is the spatial three-velocity of the matter computed by Eulerian observers in their local proper reference frame. It is easy to picture this by reference to Figure 1.1. The coordinate observer will move along with the fluid if the step in t by an amount αdt is accompanied by a shift of an amount $\vec{V}\alpha dt$ on the next hypersurface. This choice is particularly useful in spherical collapse or other inately one-dimensional problems.

A third possibility is that of the *mixed Euler–Lagrangian gauge*. In this choice, the computational grid is neither pure Eulerian nor Lagrangian. Rather, the shift vector components are chosen to simplify the solution of the field equations. This gauge is useful in hydrodynamics because it minimizes matter motion with respect to the grid and hence minimizes the associated numerical error in the hydrodynamics. In an orbiting binary, for example [21], one choice is to let [5]

$$\vec{\beta} = \vec{\Omega} \times \vec{R} + \vec{\beta}_{drag} \tag{1.76}$$

where the angular frequency Ω is chosen so that the volume averaged V^i are minimized.

Spherical geometry. In calculations of spherical collapse there are three choices. One is to choose β^r such that $B = 1$ in Eq. (1.63). This

Schwarzschild gauge is a generalization of Schwarzschild coordinates. A second possibility is to select β^r to maintain $A = B$ in Eq. (1.63). This *isotropic gauge* is the spherical version of isotropic coordinates. In this case A is like a conformal factor. It is also a special case of the minimal distortion gauge discussed below. Thirdly one can set $A = 1$ in Eq. (1.63). For this *radial gauge*, the coordinate r is the proper radial distance in the three-space.

Axisymmetry. A useful formulation for axisymmetric systems is that of the *Wilson isothermal gauge* [20]. In this application β_r and β_z are chosen to keep the dynamic three-metric in the same form as the static Weyl three-metric (1.68). This enormously simplifies the remaining Einstein equations as discussed in Chapter 6.

The extension of the isothermal gauge to the rotating case [15] is to add nonvanishing $(\beta^\phi, \gamma_{rz}, \gamma_{z\phi})$, where β^ϕ is chosen by demanding that

$$\gamma_{r\phi} - \gamma_{z\phi} = \partial_t(\gamma_{r\phi} - \gamma_{z\phi}) = 0. \tag{1.77}$$

This leaves three functions for which to solve in the three-metric

$$dl^2 = [A(dr + dz) + 2Ed\phi](dr + dz) + Dr^2 d\phi^2. \tag{1.78}$$

Another possible choice [22] is the *minimal distortion gauge* in which the shift vector is chosen to minimize the relative stretching or changing of shape (coordinate shear) when moving from one slice to the next.

This completes the general relativistic field equations in the (3+1) formalism. We still, however, must specify the detailed applications of the various gauge choices to specific problems. This is the subject of subsequent chapters.

References

[1] P. Anninos, D. Hobill, E. Seidel and L. Smarr, *Phys. Rev. Lett.*, **71**, 2851 (1993).

[2] P. Anninos, D. Hobill, E. Seidel and L. Smarr, *Phys. Rev.*, **D54**, 6544 (1996).

[3] R. Arnowitt, S. Deser and C. W. Misner, in *Gravitation*, ed. L. Witten (Wiley, New York, 1962), p. 227.

[4] C. Bona, J. Masso', E. Seidel and J. Stela, *Phys. Rev. Lett.*, **75**, 600 (1995); *Phys. Rev.*, **D56**, 3405 (1997).

[5] J. M. Bowen and J. W. York, Jr., *Phys. Rev.*, **D21**, 2047 (1980).

[6] J. Centrella and J. R. Wilson, *Astrophys. J.*, **273**, 428 (1983); *Astrophys. J. Suppl. Ser.*, **54**, 229 (1984).

[7] G. B. Cook, M. F. Huq, S. A. Klasky, M. A. Scheel, A. M. Abrahams, A. Anderson, P. Anninos, T. W. Baumgarte, N. T. Bishop, S. R. Brandt, J. C. Browne, K. Camarda, M. W. Choptuik, R. R. Correll, C. R. Evans, L. S. Finn, G. C. Fox, R. Gómez, T. Haupt, L. E. Kidder, P. Laguna, W. Landry, L. Lehner, J. Lenaghan, R. L. Marsa, J. Masso, R. A. Matzner, S. Mitra, P. Papadopoulos, M. Parashar, L. Rezzolla, M. E. Rupright, F. Saied, P. E. Saylor, E. Seidel, S. L. Shapiro, D. Shoemaker, L. Smarr, W. M. Suen, B. Szilágyi, S. A. Teukolsky, M. H. van Putten, P. Walker, J. Winicour and J. W. York, The binary black hole grand challenge alliance, *Phys. Rev. Lett.*, **80**, 2512 (1998); J. A. Font, M. Miller, W.-M. Suen and M. Tobias, *Phys. Rev.*, **D61**, 044011 (2000).

[8] C. R. Evans, Ph.D. Thesis, University of Texas (1985).

[9] E. Flanagan, *Phys. Rev. Lett.*, **82**, 1354 (1999).

[10] A. Lichnerowicz, *J. Math. Pures Appl.*, **23**, 37 (1944).

[11] G. J. Mathews and J. R. Wilson, *Phys. Rev.*, **61**, 127304 (2000).

[12] M. May and R. H. White, *Methods Comput. Phys.*, **73**, 219 (1967).

[13] C. W. Misner, K. S. Thorne and J. A. Wheeler, *Gravitation* (Freeman, San Francisco, CA, 1973).

[14] M. Shibata and T. Nakamura, *Phys. Rev.*, **D52**, 5428 (1995); M. Shibata, *Phys. Rev.*, **D60**, 104052 (1999); M. Shibata and K. Uryu, *Phys. Rev.*, **D61**, 064001 (2000); T. W. Baumgarte and S. L. Shapiro, *Phys. Rev.*, **D59**, 024002 (1999).

[15] L. Smarr, in *Sources of Gravitational Radiation*, ed. L. Smarr (Cambridge University Press, Cambridge, 1979), p. 245.

[16] L. Smarr, C. Taubes and J. R. Wilson, in *Essays in Relativity: A Festschrift for Abraham Taub*, ed. F. Tipler (Academic Press, New York, 1980), pp. 157–183.

[17] R. F. Stark and T. Piran, *Phys. Rev. Lett.*, **55**, 891 (1985).

[18] F. Usui, K. Uryu and Y. Eriguchi, *Phys. Rev.*, **D61**, 024039 (2000).

[19] S. Weinberg, *Gravitation and Cosmology*, (Wiley, New York, 1973).

[20] J. R. Wilson, in *Sources of Gravitational Radiation*, ed. L. Smarr (Cambridge University Press, Cambridge, 1979), p. 423; *Phys. Rev.*, **D66**, 084015 (2002).

[21] J. R. Wilson, G. J. Mathews and P. Marronetti, *Phys. Rev.*, **D54**, 1317 (1996).

[22] J. W. York, Jr., in *Sources of Gravitational Radiation*, ed. L. Smarr (Cambridge University Press, Cambridge, 1979), p. 83.

[23] J. W. York, Jr., in *Gravitational Radiation*, ed. N. Deruelle and T. Piran (University of Texas Press, Austin, TX, 1983), p. 147.

2

Special relativistic hydrodynamics

2.1 Perfect fluid energy momentum tensor

It is easiest to begin a study of general relativistic hydrodynamics with a discussion of special relativistic hydrodynamics in flat space. A good starting point for any hydrodynamic discussion is the energy momentum tensor. Many astrophysical and cosmological models make use of the hydrodynamics of a perfect fluid. A perfect fluid is simply one in which the velocity field at each point is such that an observer moving with this velocity sees the fluid as isotropic and nonviscous on a small enough local scale. Although we begin our discussion with a perfect fluid, we will add imperfections (viscosity, heat flow, etc.) later.

The stress tensor in Newtonian fluids is a measure of the momentum flux. In relativity (both special and general) one desires a generalization to a measure of not only momentum flux density, but also energy flux density and energy density as observed in different reference frames. In special relativity it is easy to find the form of the energy momentum tensor. Consider a pressureless cloud consisting of a local number density of n particles per unit volume with a rest mass of m_0 each. An observer in an arbitrary frame then measures a fluid mass energy density,

$$\text{mass energy density} = \sum_N \frac{nm_0}{(1 - v^2)}. \tag{2.1}$$

where one factor of $1/\sqrt{1 - v^2}$ arises from length contraction and another factor from the mass energy. In special relativity, this quantity represents the flux of mass energy across a $t = $ constant surface. The energy flux density across a spatial surface is the amount of energy transported in various directions as seen by the observer,

$$(\text{energy flux density})_i = (\text{momentum density})_i = \sum_N \frac{nm_0 v^i}{(1 - v^2)}. \tag{2.2}$$

23

This is the amount of energy transported along the ith direction per unit time. Similarly, the momentum flux along direction i from the jth component of momentum is

$$\text{momentum flux density} = \sum_N \frac{nm_0 v^i v^j}{(1-v^2)}. \tag{2.3}$$

We now denote the quantity U^μ as the relativistic four-velocity of the fluid relative to the observer,

$$U^\mu = \frac{dx^\mu}{d\tau} = W(1, V^1, V^2, V^3), \tag{2.4}$$

where V^i are the spatial velocities, dx^i/dt, and in flat Minkowski space, W is the special relativistic "gamma" factor

$$W(flat\ space) = U^t = \frac{1}{\sqrt{1 - V^i V_i}}. \tag{2.5}$$

This immediately leads to the energy momentum tensor for point particles,

$$T^{\mu\nu}_{pressureless} = \sum_N nm_0 U^\mu U^\nu. \tag{2.6}$$

Now we consider the particles to constitute a continuous fluid of rest mass energy density $\rho = nm_0$. We also divide the particle velocities into a bulk fluid velocity U^i plus isotropic velocities relative to a frame in which the fluid is at rest. The isotropic velocities then produce a pressure P, and a specific internal energy ϵ. The energy momentum tensor is then written:

$$T^{\mu\nu} = (\rho(1 + \epsilon) + P)U^\mu U^\nu + \eta^{\mu\nu} P. \tag{2.7}$$

Note that in the curved space of general relativity, W takes on a slightly different form which we deal with in subsequent chapters. This is why we distinguish it here from the usual relativistic gamma factor. The four-velocity is normalized such that

$$U_\alpha U^\alpha = -1. \tag{2.8}$$

The quantity $\eta_{\mu\nu}$ in Eq. (2.7) is the metric tensor which describes proper spacetime intervals in flat space. For example, in flat Cartesian coordinates,

$$ds^2 = \eta_{\mu\nu} dx^\mu dx^\nu, \tag{2.9}$$

with

$$\eta_{\mu\nu} = \begin{pmatrix} -1 & 0 & 0 & 0 \\ 0 & 1 & 0 & 0 \\ 0 & 0 & 1 & 0 \\ 0 & 0 & 0 & 1 \end{pmatrix}, \tag{2.10}$$

for which

$$\eta^{\alpha\gamma}\eta_{\gamma\beta} = \delta^\alpha{}_\beta. \tag{2.11}$$

2.2 Equation of motion

To describe the evolution of a relativistic fluid, it is convenient to derive equations of motion in a form that is reminiscent of the Newtonian equations of motion [24, 29, 30]. We do this by first introducing a set of Lorentz-contracted state variables. We define a coordinate baryon mass density,

$$D = W\rho. \tag{2.12}$$

Similarly, the coordinate internal energy density is written,

$$E = W\rho\epsilon. \tag{2.13}$$

We also introduce a contravariant four-space version of the coordinate three-velocity given by

$$V^\mu = \frac{U^\mu}{U^t} = (1, V^1, V^2, V^3). \tag{2.14}$$

The equation of baryon flux conservation is,

$$(\rho U^\mu)_{;\mu} = 0. \tag{2.15}$$

In Cartesian flat Minkowski space, this reduces to a special relativistic continuity equation,

$$\frac{\partial D}{\partial t} = -\frac{\partial}{\partial x^i}\left(DV^i\right), \tag{2.16}$$

which looks exactly like the Newtonian equation of baryon conservation with the Newtonian rest mass density replaced with the Lorentz-contracted mass energy density. The term on the right-hand side is an advection term. As we shall see, its accurate numerical solution requires considerable forethought.

To impose momentum conservation, we first introduce the coordinate four-momentum density,

$$S_\mu = [\rho(1 + \epsilon) + P]WU_\mu = (D + E + PW)U_\mu. \tag{2.17}$$

One can also eliminate the explicit treatment of pressure in the inertial density by introducing an index Γ from the equation of state which relates pressure and internal energy,

$$\Gamma = 1 + \frac{P}{\rho\epsilon}, \tag{2.18}$$

where Γ must be slowly varying but not necessarily constant. In this way the momentum density can be written

$$S_\mu = (D + \Gamma E)U_\mu. \tag{2.19}$$

With the above definitions we can rewrite the stress energy tensor for a perfect fluid,

$$T^\nu{}_\mu = \eta^{\lambda\nu}T_{\mu\lambda} = S_\mu V^\nu + \delta^\nu{}_\mu P, \tag{2.20}$$

where $\delta^\nu{}_\nu$ is the Kronicker delta function.

An equation for the spatial components of momentum density is derived from the vanishing of the four-divergence of the appropriate spatial terms of the energy momentum tensor,

$$T^\nu{}_{i;\nu} = 0. \tag{2.21}$$

The momentum equation in flat Cartesian coordinates then becomes

$$\frac{\partial S_i}{\partial t} + \frac{\partial}{\partial x^j}\left(S_i V^j\right) + \frac{\partial P}{\partial x^i} = 0. \tag{2.22}$$

Again this looks like the Newtonian momentum equation but with a Lorentz-contracted momentum density replacing ρv_i in the Newtonian equations. Now, in addition to the advection of momentum, we have a term (dP/dx) which accounts for the pressure force.

The energy evolution of the system is conveniently evolved using an equation for the internal energy $\rho\epsilon$ and not the total energy. One advantage of this choice is that it allows one to maintain constant entropy when desired. Also, a check for conservation of total energy or a lack thereof is a good test of numerical accuracy. Experience has shown that this is the best approach.

To obtain an equation for internal energy, we write,

$$\left(U_\mu T^{\mu\nu}\right)_{;\nu} = 0. \tag{2.23}$$

This can be reduced to an equation for internal energy density,

$$\frac{\partial(EV^\mu)}{\partial x^\mu} + P\frac{\partial U^\mu}{\partial x^\mu} = 0, \tag{2.24}$$

which becomes

$$\frac{\partial E}{\partial t} + \frac{\partial}{\partial x^i}\left(EV^i\right) + P\left[\frac{\partial W}{\partial t} + \frac{\partial}{\partial x^i}\left(WV^i\right)\right] = 0. \tag{2.25}$$

The terms containing P now describe the proper PdV work of the system. For numerical integration it is convenient to rewrite the PdV terms using the equation of state index Γ to give

$$\frac{\partial E}{\partial t} + \frac{\partial}{\partial x^i}\left(EV^i\right) + (\Gamma - 1)\frac{E}{W}\left[\frac{\partial W}{\partial t} + \frac{\partial}{\partial x^i}\left(WV^i\right)\right] = 0. \qquad (2.26)$$

Thus, the PdV contribution to $\dot{E}$ involves the product of E times slowly varying terms. They can then be easily integrated using operator splitting as described below.

2.2.1 Viscosity and heat flow

The generalization of the energy momentum tensor to an imperfect fluid includes the possibility of microscopic interactions which lead to viscosity and heat flow. To do this we follow the method of Eckart [7] described in Weinberg [28]. The stress energy tensor becomes

$$T^{\mu\nu} = (\rho(1 + \epsilon) + P)U^\mu U^\nu + \eta^{\mu\nu}P + \Sigma^{\mu\nu}, \qquad (2.27)$$

where a new viscous stress tensor $\Sigma^{\mu\nu}$ has been added which includes terms for damping and heat flow. Specifically one can write,

$$\Sigma^{\mu\nu} = -\eta h^{\mu\rho}h^{\nu\sigma}W_{\rho\sigma} - \chi\left(h^{\mu\rho}U^\mu + h^{\nu\rho}U^\mu\right)Q_\rho$$
$$-\zeta h^{\mu\nu}U^\rho{}_{;\rho}, \qquad (2.28)$$

where η is the *shear viscosity coefficient*, χ is the *heat conduction coefficient*, and ζ is the *bulk viscosity coefficient*.

The tensor $h^{\mu\nu}$ in this equation is a projection operator on the hyperplane normal to U^μ,

$$h^{\mu\nu} = g^{\mu\nu} + U^\mu U^\nu. \qquad (2.29)$$

The shear tensor $W^{\mu\nu}$ describes the velocity gradient for viscosity,

$$W_{\mu\nu} = U_{\mu;\nu} + U_{\nu;\mu} - \frac{2}{3}g_{\mu\nu}U^\sigma{}_{;\sigma}. \qquad (2.30)$$

The quantity Q_μ in Eq. (2.28) is called the *heat flow vector*,

$$Q_\mu = T_{;\mu} + TU^\nu U_{\mu;\nu}, \qquad (2.31)$$

where T is the temperature. The second term in this equation is required to keep the entropy increasing in time. It is an odd term in that it drives an acceleration of the fluid.

2.2.2 Relativistic Navier–Stokes equation

If heat flow and viscosity are added to the stress energy tensor, the
equations of motion easily generalize to the special relativistic Navier–
Stokes equation,

$$\dot{D} + \frac{\partial}{\partial x^i}\left(DV^i\right) = 0, \tag{2.32}$$

$$\dot{S}_i + \frac{\partial}{\partial x^j}\left(S_i V^j\right) + \frac{\partial P}{\partial x^i} + \frac{\partial}{\partial x^\nu}\Sigma_i^\nu = 0, \tag{2.33}$$

and

$$\dot{E} + \frac{\partial}{\partial x^i}\left(EV^i\right) + P\left(\dot{W} + \frac{\partial(WV^i)}{\partial x^i}\right) + \Sigma_\mu^\nu U_{;\nu}^\mu - \left(U^\mu \Sigma_\mu^\nu\right)_{;\nu} = 0. \tag{2.34}$$

2.2.3 Causality and Navier–Stokes

Although the relativistic Navier–Stokes equations are formally correct,
they are acausal. That is, viscous forces and heat flow can propagate at
arbitrary speeds. To avoid this possible problem it is common to introduce
flux limiters. In the fluid frame these equations reduce to the form of a
diffusion equation, $\dot{A} = D\partial^2 A/\partial x^2$. Hence, the same flux limiter methods
which have been developed to limit diffusion in Newtonian hydrodynamics
can be applied here.

   That is, for either viscosity or heat flow, one identifies the flux of
momentum Φ_S and/or heat Φ_H and imposes the following requirements:

$$|\Phi_S| < (\rho + \rho\epsilon + P)c_s, \tag{2.35}$$
$$|\Phi_H| < \rho\epsilon_{th}c_s, \tag{2.36}$$

where c_s is the speed of sound and ϵ_{th} is the thermal component of the
internal energy per unit mass. To achieve these constraints one goes back
to the fluid frame and identifies the momentum flux with $\eta(\partial U^i/\partial x^i)$ and
the heat flux in Eq. (2.28) with $\kappa(\partial T/\partial x^i)$. These are then replaced by

$$\Phi_S = \eta(\partial U^i/\partial x^i)/\left(1 + \frac{\eta|(\partial U^i/\partial x^i)|}{(\rho + \rho\epsilon + P)c_s}\right), \tag{2.37}$$

and the heat flux with

$$\Phi_H = \kappa(\partial T/\partial x^i)/\left(1 + \frac{\kappa|(\partial T/\partial x^i)|}{\rho\epsilon_{th}c_s}\right). \tag{2.38}$$

In Chapter 5 we will discuss how to design flux limiters to obtain good
quantitative numerical results.

2.3 Coordinate systems

There are two basic choices for flat space coordinate systems. The equations above are written in an *Eulerian* scheme. That is, a fixed grid is defined and the density and other state variables become functions of position on this grid, i.e we write $\rho(x^\mu)$. The other possible scheme is a *Lagrangian* representation. Here, the density and coordinates become functions of mass and time, i.e. we have $\rho(m, t)$, $x^i(m, t)$. The advantage of a Lagrangian scheme is apparent when the fluid velocities obey a simple functional behavior (e.g. spherical collapse or expansion). By choosing the appropriate mass coordinate, baryon conservation can be guaranteed. However, if the fluid motion is complicated, for example by the existence of vortices or shear, there is no simple relation between initial and final mass points, and an Eulerian scheme is better.

It has often proven to be most useful to consider a mixing of the advantages of both the Eulerian and Lagrangian schemes. That is, if at least part of the fluid motion can be described by a simple fluid grid velocity V_g, then a coordinate transformation, $dx^i \to dx^i + V_g^i dt$ produces,

$$\frac{\partial D}{\partial t} + \nabla \cdot \left(D(\vec{V} - \vec{V}_g) \right) + D\nabla \cdot \vec{V}_g = 0, \tag{2.39}$$

$$\frac{\partial E}{\partial t} + \nabla \cdot \left(E(\vec{V} - \vec{V}_g) \right) + E\nabla \cdot \vec{V}_g$$

$$+ P\left[\frac{\partial W}{\partial t} + W\nabla \cdot V_g + W\nabla \cdot \left(\vec{V} - \vec{V}_g \right) \right] = 0, \tag{2.40}$$

$$\frac{\partial S_i}{\partial t} + \nabla \cdot \left(S_i(\vec{V} - \vec{V}_g) \right) + S_i \nabla \cdot \vec{V}_g + \frac{\partial P}{\partial x^i} = 0. \tag{2.41}$$

Note that for $\vec{V}_g = 0$ these reduce to the Eulerian scheme. For $\vec{V}_g = \vec{V}$, the above reduces to a Lagrangian scheme.

As written, these equations are for fluid flow in which each particle retains its initial entropy. One could also easily add an external energy source $\dot{E}_{external}$ to the energy equation. Also, if we wish to describe shock waves we must add terms for viscosity and heat flow as described in the previous section. We will return to these terms shortly. First, however, we should say a few words about how one actually reduces the above equations to something that can be evolved numerically on a computer.

2.4 Difference equations: generalities

A number of schemes have been introduced to evolve hydrodynamic equations [15]. At the heart of all approaches is the treatment of advection terms which appear in equations such as (2.16), (2.22), and (2.25). Some of the modern algorithms like the Riemann solvers [20] have found wide application for a number of problems for both Newtonian and relativistic hydrodynamics. The modern artificial viscosity (AV) shock capturing method described herein is based upon the numerical techniques developed during the past decade for applications to Newtonian hydrodynamics at Lawrence Livermore National Laboratory (LLNL). Although there is a general conception (cf. [15]) that the modern Riemann methods are superior to artificial viscosity, this conclusion is based upon comparisons with older versions of the AV method as applied in the 1980s [4, 12, 13].

The algorithms presented here are based upon a relativistic generalization of the modern artificial viscosity. This method has evolved a long way since the old Von Neuman–Richtmer [27] artificial viscosity which needed many zones to achieve an accurate representation of a shock wave and also was dissipative when not appropriate. All comparisons to AV methods in use prior to those described in [16] are irrelevant. The scheme described herein is based upon the more recent AV method which we consider to be preferable to many Riemann solvers currently in use.

2.4.1 Artificial viscosity versus Riemann

Extensive comparisons have been made [14] at LLNL in recent years between the modern AV and Riemann methods. For one-dimensional test problems, the Riemann solvers are indeed slightly superior. In the tests cited above [14] some multi-dimensional flows were found to be slightly better represented by AV than by the Riemann method. The Riemann method, however, is more easily amenable to adaptive mesh refinement techniques [2, 3]. All Riemann solvers, however, are more dissipative of energy in smooth flow than the modern AV methods. For problems that involve strong energy flow and/or real viscosity (e.g. supernovae), the Riemann solvers become complicated if heat flow, etc. are incorporated into the characteristics. Furthermore, for three-dimensional calculations the operator splitting of the Riemann solver usually involves three separate one-dimensional sweeps across the grid. This can cause a degradation in accuracy for three-dimensional applications.

Figure 2.1 was prepared for us by Peter Anninos [1]. It illustrates a comparison between the three-dimensional structure for a simple Newtonian neutron star evaluated using both the Riemann total variation diminishing (TDV) method [11] and an AV method for the same number of zones. For a one-dimensional test problem, the TDV solver was as good as or better than the AV method. However, this illustrates two problems when the TDV is applied in three dimensions: the central

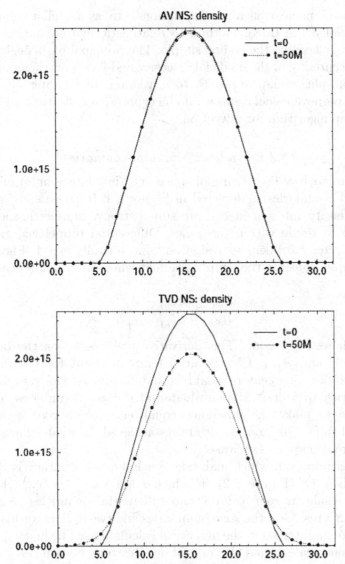

Fig. 2.1. Radial profile of a three-dimensional Newtonian neutron star calculated [1] using both artificial viscosity (AV) and a TDV method.

density is too low and the surface is poorly resolved. Both of these effects can be traced to the tendency of Riemann solvers to introduce excess dissipation. If the characteristics of the TDV method could be solved in a true implicit three-dimensional manner, this problem might be eliminated. We also note, however, that the TDV methods make less efficient use of computer resources.

We believe that the modern AV technique described herein is robust for problems that involve multi-dimensional hydrodynamics, heat flow, real viscosity, nuclear burning, detonations, etc. as in stellar collapse. We have extended the LLNL method to relativistic hydrodynamics as best we can while keeping the coding simple. The principal parts neglected are the time centering of the artificial viscosity and incorporating an effective Lagrangian-plus-remap approach to advection. In Chapter 5, however, for the supernova calculations we do incorporate an effective Lagrangian-plus-remap algorithm for advection.

2.4.2 Finite difference preliminaries

One begins with a discretizing of space and time into a finite number of zones and boundaries as depicted in Figure 2.2. Intrinsic state variables like the density, internal energy, pressure, entropy, and metric coefficients are taken to reside within the zones. Differential properties, gradients, velocities, etc. are taken to reside on zone boundaries (cf. Figure 2.2). For example, consider the simple gradient when finite differenced in one direction,

$$\frac{\partial A}{\partial x} \rightarrow \frac{A_{j+1} - A_j}{x_{j+1} - x_j}, \tag{2.42}$$

where j labels each zone. This derivative now resides on the boundary between A_j and A_{j+1}. Clearly, an attempt to treat the derivatives as located at the zone centers would be inaccurate. At the very least one must average over derivatives evaluated at the two boundaries. The goal should be to make the difference equations second-order accurate in space and time. This goal is not always achieved. In what follows we will note discrepancies as they arise.

A better approach which maintains second-order accuracy is the leap-frog method (cf. Figure 2.2) in which state variables (e.g. $A_{i+1/2} \equiv D, E, P$) reside in zone centers and differential quantities (e.g. $B_i \equiv V_i, U_i, W, S_i$) reside on the zone boundaries and two different grids ($x_{j+1/2}$ and x_j) are maintained in the numerical calculations. The first $x_{j+1/2}$ grid denotes zone centers, and the x_j grid denotes zone boundaries

$$
\begin{array}{llllll}
B_{i-1} & & B_i & & B_{i+1} & \\
| & A_{i-1/2} & | & A_{i+1/2} & | & \\
x_{i-1} & x_{i-1/2} & x_i & x_{i+1/2} & x_{i+1} &
\end{array}
$$

The same staggering is made in the time steps which we denote with a superscript n, i.e.

$$
\begin{array}{llllll}
B^{n-1} & & B^n & & B^{n+1} & \\
| & A^{n-1/2} & | & A^{n+1/2} & | & \\
t^{n-1} & t^{n-1/2} & t^n & t^{n+1/2} & t^{n+1} &
\end{array}
$$

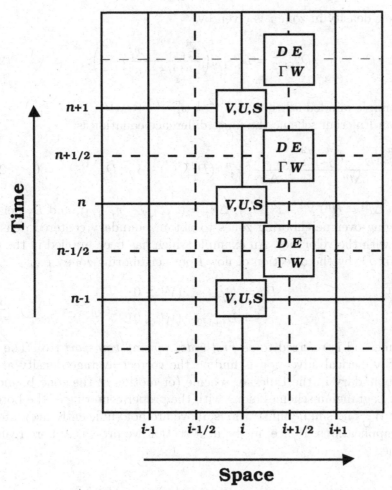

Fig. 2.2. Schematic illustration of the spacetime centering of various variables in the leap-frog method.

As an illustration, consider the simple planar one-dimensional continuity equation

$$\dot{D} + \frac{\partial}{\partial x}(DV) = 0. \tag{2.43}$$

In a Lagrangian scheme the mass conserving formulation of this equation is achieved by simply following the change in volume (here length) between the zone boundaries from one time step to the next. That is, if we introduce x_i^{n+1} as the new zone boundaries, then in one dimension the new density in zone i is given by,

$$D_{i+1/2}^{n+1} = D_{i+1/2}^n \left(\frac{x_{i+1}^{n-1/2} - x_i^{n-1/2}}{x_{i+1}^{n+1/2} - x_i^{n+1/2}} \right). \tag{2.44}$$

In a simple minded scheme, $x_i^{n+1/2} = x_i^{n-1/2} + \Delta t V^i$.

In an Eulerian scheme the finite difference equation is

$$\frac{D_{j+1/2}^{n+1/2} - D_{j+1/2}^{n-1/2}}{\Delta t^n} + \frac{1}{\Delta x_j} \left(V_{j+1}^n (\tilde{D}_{j+1}^{n+1/2}) - V_j^n (\tilde{D}_j^{n+1/2}) \right) = 0, \tag{2.45}$$

where $\Delta t^n = (t^{n+1/2} - t^{n-1/2})$, $\Delta x_j = (x_{j+1/2} - x_{j-1/2})$, and $\tilde{D}_j$ denotes averaging over neighboring zones to obtain boundary centered densities to go with the velocities. An example which has been applied in the past is to fix $\tilde{D}_j$ by the direction of flow from neighboring zones, e.g.

$$\begin{aligned}
\tilde{D}_j &= D_{j-1/2} \quad (V^j > 0) \\
&= D_{j+1/2} \quad (V^j < 0).
\end{aligned} \tag{2.46}$$

This method is referred to as *first-order upwind transport* [13]. The real art of numerical advection is finding the correct average density at the zone boundary in the Eulerian scheme (or motion of the zone boundary in the Lagrangian scheme) along with the proper centering of the Lorentz factor W. The simple illustrations above are not sufficiently accurate for most applications. Hence, in the next section we present a more realistic scheme.

2.4.3 Relativistic hydrodynamics in one dimension

In this section we give the difference equations in a linear one-dimensional Eulerian mode. The extension to general relativity and higher dimensions is presented in the next chapter. A method which makes use of the Lagrangian mode (in a hybrid scheme) will be given in the chapter on supernova collapse models. A one-dimensional hydrodynamics code which incorporates the features described herein along with codes for generating test problems is available by request from the authors (email: gmathews@nd.edu).

We begin by specifying two staggered positional grids. The points $x_i \equiv x_a^i$ are centered on the boundaries between zones and define the positions at which velocity quantities, S_i, U_i, V^i, β^i, are centered. (Note that in the following the subscripts and superscripts denote zone positions in the one-dimensional numerical grid and not spatial dimensions.)

We use the zone-centered grid $x_{i+1/2} \equiv x_b^i$ to denote positions for matter variables, i.e. D_i, E_i, etc. Increments in x are labeled by $dx_a^i = x_b^i - x_b^{i-1}$ for the boundary-centered x_i positions and $dx_b^i = x_a^{i+1} - x_a^i$. The variables D_i, E_i, Γ_i, and P_i are treated as centered in time. They are constructed so to be one half time step away from the time centering of velocity quantities, S_i, U_i, etc. The Lorentz factors, W_i, are required as both zone- and boundary-centered quantities. However, they are derived from velocities and would more naturally be defined only on the zone boundaries. They must continually be averaged from the zone boundaries, hence the centering of W^i is not as easy to maintain.

To generalize the treatment of planar, cylindrical, and spherical problems with a single algorithm, it is necessary also to specify areas and volumes on the grid. Hence, for the boundary-centered areas A_a^i and volumes Vol_a^i we write,

$$
A_a^i = \begin{array}{ll}
1 & \text{planar} \\
2\pi x_a^i & \text{cylindrical} \\
4\pi (x_a^i)^2 & \text{spherical,}
\end{array} \tag{2.47}
$$

$$
Vol_a^i = \begin{array}{ll}
x_b^i - x_b^{i-1} & \text{planar} \\
\pi[(x_b^i)^2 - (x_b^{i-1})^2] & \text{cylindrical} \\
4\pi[(x_b^i)^3 - (x_b^{i-1})^3]/3 & \text{spherical.}
\end{array} \tag{2.48}
$$

Similarly, the zone-centered areas and volumes are,

$$
A_b^i = \begin{array}{ll}
1 & \text{planar} \\
2\pi x_a^i & \text{cylindrical} \\
4\pi (x_a^i)^2 & \text{spherical,}
\end{array} \tag{2.49}
$$

$$
Vol_b^i = \begin{array}{ll}
x_a^i - x_a^{i-1} & \text{planar} \\
\pi[(x_a^i)^2 - (x_a^{i-1})^2] & \text{cylindrical} \\
4\pi[(x_a^i)^3 - (x_a^{i-1})^3]/3 & \text{spherical.}
\end{array} \tag{2.50}
$$

2.4.4 Operator splitting

We make extensive use of operator splitting. That is, the various coupled equations are solved successively while ignoring the other equations at each time step. The separate steps by which the system is advanced in time are as follows:

1. pressure acceleration
2. viscosity
3. velocities U, V and W
4. pressure PdV work on fluid
5. advection of state variables
6. velocities again
7. pressure PdV work again
8. time step dt calculation
9. grid update
10. output and post processing when appropriate.

The equation-of-state routine is called after each step in which the state variables are changed, i.e. steps 2, 4, 5 and 7. The order of these steps is important and should be performed as listed above. This splitting insures that variables are optimally updated.

2.4.5 Time step calculation

The time steps dt_n are taken as the minimum of the time step as determined by several conditions. Each condition is also multiplied by a number less than one to accommodate the nonlinear nature of the equations.

The first condition is known as the Courant condition. It is based upon a search over all zones i for the zone with the minimum sound crossing time, i.e.

$$dt_1 = \min\left(dx_b^i/c_s^i\right), \tag{2.51}$$

where c_s^i is the sound speed in the ith zone.

The speed of sound is found by rewriting the Eulerian equations of motion into a wave equation for small perturbations in the density. The Newtonian speed of sound is given by the variation of pressure with density. In relativity the wave speed is given instead by the adiabatic derivative of the pressure with respect to the relativistic inertial density,

$$c_s^2 = \frac{\partial P}{\partial[\rho(1 + \epsilon) + P]}. \tag{2.52}$$

For an equation of state such that P_i in zone i may be represented locally by a local adiabatic index Γ_i, we have

$$P_i = (\Gamma_i - 1)E_i/W_i. \tag{2.53}$$

The local sound speed in zone i then becomes,

$$c_s = \sqrt{\frac{\Gamma_i(\Gamma_i - 1)E_i}{D_i + \Gamma_i E_i}}. \tag{2.54}$$

For $\Gamma > 1$ the wave speed increases with density if adiabaticity is assumed. This is because the specific internal energy ϵ increases with density. Thus, waves may steepen as they propagate.

The second condition is a search for the zone with minimum three-velocity transport time, defined as the time for material to flow across a zone,

$$dt_2 = \min \left(dx_a^i / |V^i| \right). \tag{2.55}$$

This constraint is introduced to ensure stability and accuracy in the numerical advection calculation.

The third condition is introduced to maintain stability of the artificial viscosity algorithm. As described below, the viscous equations are analogous to a diffusion equation in four-velocity with a diffusion coefficient $D \approx k_1 dx^i |\delta U^i|$, where $\delta U^i \equiv U^{i+1} - U^i$. We then can define a minimum viscous diffusion time across a zone derived from the stability condition for explicit diffusion equations,

$$dt_3 = \frac{1}{4} \min \left(W^i dx_b^i / |\delta U^i| \right). \tag{2.56}$$

The time step dt is then assigned to be some fraction of the minimum of these three conditions,

$$dt = k \min \left(dt_1, dt_2, dt_3 \right), \tag{2.57}$$

where typically $k \approx 1/2$. Obviously, smaller values for k increase the accuracy of the calculation. Also, since the equations are nonlinear, sometimes violent systems require a smaller value for k.

2.4.6 Artificial viscosity

The nonlinearity of the equations of hydrodynamics can lead to the formation of shocks. It is difficult to depict accurately a numerical shock. The scale of shocks is very small for most astrophysical situations,

$$L_{shock} \approx \lambda \frac{|\Delta v_s|}{|v_s|}, \tag{2.58}$$

where λ is the mean free path in the medium, v_s is the shock velocity and Δv_s is the difference between the post-shock and pre-shock velocities. Since we almost always deal with zone sizes much larger than λ, it is usually impossible to resolve a shock. Furthermore, the large gradients in velocity as a shock passes lead to difficulty in the numerical advection scheme. An unphysical oscillation in intrinsic properties results, as we shall show in some test problems later in the chapter. Thus, some method is desired to "smooth out" shock fronts over several zones.

At the end of this chapter we give an explicit example of real viscosity for a special relativistic fluid in heavy-ion nuclear collisions. As we shall see, real physical viscosity spreads out a shock front. In analogy with physical viscosity, therefore, an artificial viscosity is introduced for numerical calculations which becomes large when a velocity gradient is large and small otherwise. Its purpose is to reproduce the shock structure with a minimum number of zones.

In the original form as introduced by Von Neuman and Richtmer (1950) a viscosity coefficient was postulated which has more or less the desired properties,

$$\eta = k\rho|\Delta V|\Delta x, \tag{2.59}$$

where $\Delta V = V_{j+1} - V_j$ is the change in velocity across a zone and $\Delta x = x_{j+1} - x_j$ is the zone width. Over the years more sophisticated forms have been proposed. This viscosity coefficient is then incorporated into the hydrodynamic equations to modify the momentum energy evolution. For relativistic applications one must be careful to make the proper generalization from three-velocity to four-velocity and include the Lorentz factors. The method described below is the one preferred by the authors. As we shall see it provides good accuracy for most test problems.

In this algorithm, the artificial viscosity coefficient becomes

$$\eta = \sigma[k_1\delta U - k_2 c_s], \tag{2.60}$$

where σ is the inertial density and δU is an appropriately evaluated change in four-velocity across two zones. The implementation of the viscous stress tensor in one spatial dimension reduces to the evaluation of a pressure-like quantity Q for the momentum equation,

$$Q \equiv \sigma\delta U(k_1\delta U - k_2 c_s), \tag{2.61}$$

and a kinetic energy loss term involving

$$C \equiv k_3\bar{\sigma}\frac{\delta U}{W}|(V - V_g)|, \tag{2.62}$$

where V_g is a grid velocity. Typical values for the viscosity coefficients are:

$$k_1 = (\Gamma + 1)/2, \qquad k_2 = 1, \qquad k_3 = 0.25. \tag{2.63}$$

Obviously these have the desired property of large artificial viscosity when the velocity change δU is large. The trick is to center properly all of the quantities in the above expressions and avoid any artificial stress when none is required.

2.4.7 Realistic artificial viscosity in one dimension

In a realistic calculation the evaluation of the artificial viscosities is quite complicated. Here, we sketch a one-dimensional algorithm to compute artificial viscosity. A FORTRAN code which contains the basic algorithm described herein is available by request from the authors. As we shall see in the test problems described later in this chapter, this method is able to maintain good accuracy with only two to three zones in the shocks.

First, the velocities at the zone boundaries U_i and V_i and the Lorentz factor W_i are updated from the existing momentum density. We have,

$$U_i = \frac{S_i dx_a^i}{\frac{1}{2}(\sigma_{i-1} dx_b^{i-1} + \sigma_{i-1} dx_b^{i-1})}, \tag{2.64}$$

where the inertial coordinate density σ_i is

$$\sigma_i = D_i + \Gamma_i E_i. \tag{2.65}$$

The three-velocity is then

$$V_i = U_i / \sqrt{1 + U_i^2}, \tag{2.66}$$

the boundary-centered Lorentz factor is

$$W_i^a = \sqrt{1 + U_i^2}, \tag{2.67}$$

and the zone-centered Lorentz factor is taken to be

$$W_i^b = \sqrt{1 + (U_i^2 + U_{i+1}^2)/2}. \tag{2.68}$$

Next, velocity differences are determined for the Q calculation. A difference in the four-velocity at each boundary is determined from the minimum and maximum four-velocities in the vicinity of a zone,

$$U_{min} = \min(U_{i-1}, U_i, U_{i+1}), \tag{2.69}$$

$$U_{max} = \max(U_{i-1}, U_i, U_{i+1}). \tag{2.70}$$

From this a minimum derivative is determined

$$\nabla U_i = \min(U_{max} - U_i, U_i - U_{min})/\Delta x, \tag{2.71}$$

and a monotonicity condition is imposed on the derivative

$$\text{if} \quad (U_{i+1} - U_i)(U_i - U_{i-1}) < 0 \quad \text{then} \quad \nabla U_i = 0. \tag{2.72}$$

Then, we determine the average derivative at zone i,

$$\nabla\tilde{U} = (U_{i+1} - U_{i-1})/2\Delta x. \tag{2.73}$$

The actual derivative in U is taken as the minimum of this quantity or the minimum derivative across a single zone, i.e.

$$\nabla\bar{U} = \nabla\tilde{U}\min(1.0, \nabla U/|\nabla\tilde{U}|). \tag{2.74}$$

Having now obtained a minimum monotonic derivative of the four-velocity, this is now used to compute the change in four-velocity δU,

$$\delta U_i = \left[(U_{i+1} - U_i) - (1/2)\Delta t(\nabla\tilde{U}_i + \nabla\tilde{U}_{i+1})\right]/W_i^b. \tag{2.75}$$

Monotonicity is then imposed,

$$\text{if} \quad \left[\delta U_i(U_{i+1} - U_i) < 0\right], \tag{2.76}$$

then δU_i is restricted to be negative,

$$\delta U_i = \min\left(0, \delta U_i\right). \tag{2.77}$$

With δU thus constructed, we then have the spatial component of the artificial viscous stress

$$Q_i = \sigma_i\delta U_i(k_1\delta U_i - k_2 c_s^i), \tag{2.78}$$

where k_1 and k_2 are viscosity coefficients.

The momentum and energy equations can then be updated

$$S_i = S_i - \frac{Q_i - Q_{i-1}}{\Delta x}\Delta t, \tag{2.79}$$

and new velocities determined as above.

In the advection the momentum density is spread out somewhat. This leads to a loss of kinetic energy. The quantity C_i is introduced to compensate for the kinetic energy loss by increasing the internal energy as follows:

$$C_i = k_3\bar{\sigma}_i\frac{\delta U^i}{W}|(V^i + V^{i+1} - V_g^i - V_g^{i+1})/2|, \tag{2.80}$$

where V_g^i is the local grid velocity. Then, the internal energy can be updated

$$E_i = E_i - (Q_i - C_i)\left(\frac{(U^{i+1} - U^i)\Delta t}{\Delta x}\right). \tag{2.81}$$

This completes the artificial viscosity calculation. Note, however, that Q_i was evaluated using a combination of σ_i and U_i. The time centering of these two quantities differs by half a time step. Hence, the error in the artificial viscosity grows linearly with dt. Typically, with a strong shock and an allowably large dt, the error in compression across a shock front is $\sim$1–2%. To achieve higher accuracy either a small time step or a time-centered artificial stress Q_i should be used. In the Newtonian limit, properly time-centered artificial stresses have been developed for the method presented here. These reduce the shock compression error to $\sim$0.1%. For relativistic hydrodynamics, however, it is necessary also to time center W in a consistent way. The complete second-order differencing of Q and W in time has not yet been achieved.

Another way to increase numerical accuracy is to use reduced time steps in regions close to shocks. One can run multiple time steps in those regions in such a way as to keep them in time step with the other zones. In the chapter on supernova numerical methods we give an example of such a reduction in time steps.

2.4.8 Equation of state

Having reevaluated the internal energy in the artificial viscosity section, we now wish to reevaluate the pressure and temperature. The equation of state (EOS) determines P and temperature T given ρ and E. Most astrophysical processes involve gases, hence we represent the pressure by

$$P = (\Gamma - 1)E/W. \tag{2.82}$$

In the polytropic limit Γ is a constant, typically $4/3 < \Gamma < 2$. In realistic calculations, Γ is a slowly varying function of $\rho = D/W$ and $\epsilon = E/\rho W$. Hence, tables or analytic functions are often used. In the same spirit, the temperature T is determined when desired by assigning a heat capacity from the EOS defined by

$$C_V = \frac{\epsilon}{T}, \tag{2.83}$$

and assuming that C_V does not change significantly during a computational cycle.

2.4.9 Velocity calculation

Having updated the momentum density in the artificial viscosity section as well as the state variables in the equation of state, it becomes necessary to reevaluate the four-velocity,

$$U_i = S_i/\bar{\sigma}_i, \tag{2.84}$$

where $\bar{\sigma}_i$ is the inertial density, $\sigma \equiv D + \Gamma E$, averaged at the zone boundary,

$$\bar{\sigma}_i = (\sigma_i dx_b^i + \sigma_{i-1} dx_b^{i-1})/2dx_a^i. \tag{2.85}$$

In this step we also evaluate the new Lorentz factors W along with changes in the coordinate energy density E due to changes in W and the advective part of W.

The three-velocity and zone-centered Lorentz factors are also reevaluated,

$$V^i = U_i/W_a^i, \tag{2.86}$$

where W_a^i is the boundary-centered Lorentz factor,

$$W_a^i = \sqrt{1 + U_i^2}. \tag{2.87}$$

The zone-centered W factors are then

$$W_b^i = \sqrt{1 + (U_i^2 + U_{i+1}^2)/2}. \tag{2.88}$$

2.4.10 Pressure work

With the velocity updated it is good next to evaluate the pressure terms in the energy equation (2.25), $P[\frac{\partial W}{\partial t} + \frac{\partial}{\partial x^i}(WV^i)]$.

In this step it is assumed that the Γ_i remain constant through the time step. One can then rewrite the pressure as $P = (1 - \Gamma)E/W$. In this way the pressure terms in the dynamical equations can be integrated and added as an update to the state variables.

First the $\partial W/\partial t$ term can be evaluated:

$$E_i = E_i(W_b^{i,old}/W_b^{i,new})^{(\Gamma-1)}, \tag{2.89}$$

where the superscripts *new* and *old* are used to denote the new and previous values of W respectively from the velocity calculation.

Next, the $V\nabla W$ term is evaluated and the energies updated,

$$V\nabla W_b^i = \frac{1}{2}(V^i + V^{i+1} - V_g^i - V_g^{i+1})(W_a^{i+1} - W_a^i)/dx_b^i, \tag{2.90}$$

$$E_i(t + dt) = E_i(t)\exp\left((\Gamma_i - 1)\frac{V\nabla W_b^i}{W_b^i}dt\right). \tag{2.91}$$

Finally, the $WP\nabla \cdot V$ term can be evaluated. The divergence of the three-velocity is given by

$$\nabla \cdot V^i = \left(A_a^{i+1}V^{i+1} - A_a^i V^i\right)/Vol_b^i. \tag{2.92}$$

We then have,

$$E_i(t + dt) = E_i(t) \exp \left\{ (\Gamma_i - 1) \frac{\nabla \cdot V^i}{W_b^i} dt \right\}.$$ (2.93)

2.4.11 Grid update

As mentioned in Section 2.3, greater accuracy can be maintained by choosing coordinates which follow the average motion of the fluid while still maintaining the generality of the Eulerian approach. This is particularly true in multi-dimensional problems. For the one-dimensional test problems it is worthwhile to define several coordinate motions in addition to the Eulerian limit ($V_g = 0$). One is uniform flow for which $V_g^i =$ constant. Another is homologous motion (analogous to Hubble flow, i.e. $V_g \propto r$). For this case one simply sets the grid velocities and new coordinate radii as

$$V_g^i = V^{imax} (r^i / r^{imax}),$$ (2.94)

where V^i is the total fluid three-velocity and i_{max} is the last zone of the grid. Alternatively, one can select the Lagrangian limit

$$V_g^i = V^i.$$ (2.95)

Having selected a grid velocity at each zone, the coordinates are next updated:

$$x_a^i \rightarrow x_a^i + V_g^i dt,$$ (2.96)

$$dx_b^i = x_a^{i+1} - x_a^i,$$ (2.97)

$$x_b^i \rightarrow x_a^i + \frac{1}{2} dx_b^i,$$ (2.98)

$$dx_a^i = x_b^i - x_b^{i-1}.$$ (2.99)

Similarly, the volumes and areas are updated according to Eqs. (2.48) and (2.50). Finally, the state variables can be updated,

$$D_i \rightarrow D_i \left(\frac{Vol_b^{old}}{Vol_b^{new}} \right),$$ (2.100)

$$E_i \rightarrow E_i \left(\frac{Vol_b^{old}}{Vol_b^{new}} \right)^{\Gamma},$$ (2.101)

$$S_i \rightarrow S_i \left(\frac{Vol_a^{old}}{Vol_a^{new}} \right).$$ (2.102)

Note that the Γ factor in Eq. (2.101) arises from combining the $E\nabla \cdot V_g$ term and the $PW\nabla \cdot V_g$ term from Eq. (2.40).

2.4.12 Pressure acceleration

The dP/dx pressure force term is applied to the spatial component of the relativistic four-momentum density in a straightforward way,

$$S_i(t + dt) = S_i(t) - dt\frac{(P_i - P_{i-1})}{dx_a^i}.\tag{2.103}$$

After this step, U_i and V_i are updated based upon the new values of E_i D_i, and S_i from the pressure and grid calculations just completed.

2.4.13 Advection

The calculation of the advection of the matter variables, i.e. $\partial(DV)/\partial x$, $\partial(EV)/\partial x$, $\partial(SV)/\partial x$, is probably the trickiest part of the whole simulation. One must be very careful. Otherwise instabilities and/or inaccuracies will arise.

2.4.14 Density advection

We consider the advection of coordinate density D first. The treatments of energy and momentum advection are quite similar. The key to treating advection is to convert the differential equation of the form

$$\frac{\partial D}{\partial t} = -\frac{\partial(DV)}{\partial x}\tag{2.104}$$

to a properly zone-centered mass-conserving difference equation. Numerous schemes have been proposed to find the best flux between the zones (cf. [13]). The one described herein is the one currently favored by the authors. It uses the concept of *monotonicity* introduced by Roe [20] and Van Leer [26]. To conserve mass it is best to deal with the transferred particle mass ΔM_D^i into zone i from the zone below, so that the finite differenced advection term becomes

$$D^i(t + \delta t) = D^i(t) - (\Delta M_D^{i+1} - \Delta M_D^i)/Vol_b^i.\tag{2.105}$$

The transferred mass is written in terms of an optimum interpolated density (for mass flux) at the zone boundary during the time step,

$$\Delta M_D^i = \bar{D}_f^i A_a^i (V^i - V_g^i)dt.\tag{2.106}$$

The interpolated flux density passing through the boundary, in turn, is found from the best interpolated density gradient at the boundary, i.e.

$$\text{if} \quad (V^i - V_g) > 0, \quad \bar{D}_f^i = D^{i-1} + \frac{1}{2}\nabla\tilde{D}^{i-1}[dx_b^{i-1} - (V^i - V_g^i)dt],$$
$$\tag{2.107}$$

$$\text{if} \quad (V^i - V_g) < 0, \quad \bar{D}^i_f = D^i - \frac{1}{2}\nabla\tilde{D}^i[dx^i_b + (V^i - V^i_g)dt]. \quad (2.108)$$

The content of Eqs. (2.107) and (2.108) is simple. The quantity $\nabla\tilde{D}dx$ is just the density difference at the boundary at the beginning of a time step. At the end of a time step the zone transferring mass will decrease its density by an amount $\nabla\tilde{D}Vdt$. Hence, Eqs. (2.107) and (2.108) give us a guess of the density passing through the zone boundary midway through the time step.

The entire trick to this method then reduces to finding the best average density gradient $\nabla\tilde{D}^i$ at each zone boundary. This becomes slightly complicated when the density gradient is varying rapidly across a few zones (i.e. as when a shock is passing). If one chooses a gradient which is too large then too much mass will be transferred during a step and the density will oscillate. To avoid this one chooses a gradient which best preserves the monotonicity of density as a shock wave passes.

To calculate the density gradient one begins with a simple two-point Lagrange interpolation of the average density at the zone boundary,

$$D^i_a = \frac{D^i dx^{i+1}_b + D^{i+1} dx^i_b}{dx^i_b + dx^{i+1}_b}. \quad (2.109)$$

This may or may not be used to determine the density gradient, but it is always used to establish the sign of the density gradient at each boundary.

Next, maximum and minimum densities around each zone are identified

$$\begin{aligned} D_{min} &= \min\{D^{i-1}, D^i, D^{i+1}\}, \\ D_{max} &= \max\{D^{i-1}, D^i, D^{i+1}\}. \end{aligned} \quad (2.110)$$

Several possible density differences are considered for the gradient calculation. The optimum is chosen from the minimum of

$$\Delta D_{min} \equiv 2\min\{D_{max} - D^i, D^i - D_{min}\}, \quad (2.111)$$

and

$$\Delta D_{max} = \max(D_{max} - D^i, D^i - D_{min}, |\Delta D_a|), \quad (2.112)$$

where

$$\Delta D_a = D^i_a - D^{i-1}_a. \quad (2.113)$$

Thus, the optimum density gradient is then

$$\nabla\tilde{D}^i = \frac{\min(2\Delta D_{min}, \Delta D_{max})}{dx^i_b}\frac{\Delta D_a}{|\Delta D_a|}. \quad (2.114)$$

The average boundary density for $\bar{D}_f^i$ is then evaluated from Eqs. (2.107) and (2.108). The mass transferred through the boundary ΔM^i is then given from Eq. (2.106), and the coordinate density can be updated from Eq. (2.105).

2.4.15 Energy advection

The above steps are repeated for the coordinate internal energy density E which is updated from the transferred internal mass energy ΔM_E^i,

$$E_i(t + \Delta t) = E_i(t) - (\Delta M_E^{i+1} - \Delta M_E^i)/Vol_b^i, \qquad (2.115)$$

where

$$\Delta M_E^i = \bar{E}_f^i A_a^i (V^i - V_g^i) dt, \qquad (2.116)$$

and as before

$$\text{if} \quad (V^i - V_g) > 0, \quad \bar{E}_f^i = E^{i-1} + \frac{1}{2} \nabla \tilde{E}^{i-1} [dx_b^{i-1} - (V^i - V_g^i) dt],$$

$$(2.117)$$

$$\text{if} \quad (V^i - V_g) < 0, \quad \bar{E}_f^i = E^i - \frac{1}{2} \nabla \tilde{E}^i [dx_b^i + (V^i - V_g^i) dt].$$

By repeating the steps leading up to Eq. (2.114), the optimum energy density gradient is then

$$\nabla \tilde{E}^i = \frac{\min(2\Delta E_{min}, \Delta E_{max})}{dx_b^i} \frac{\Delta E_a}{|\Delta E_a|}. \qquad (2.118)$$

2.4.16 Momentum advection

From the transferred mass in particle density ΔM_D^i and internal energy ΔM_E^i above, one then can construct an amount of transferred inertial mass ΔM_I^i for the momentum advection,

$$\Delta M_I^i = \Delta M_D^i + \Delta M_E^i (\Gamma_i + \Gamma_{i-1})/2, \qquad (2.119)$$

to be used in the evaluation of the momentum flux.

The momentum advection algorithm is based upon using the flux of inertial density defined above, together with an interpolated four-velocity U. Since the momentum density is centered on the zone boundaries, the momentum flux must be defined in zone centers. Hence, we begin by evaluating zone-centered velocities and fluxes of inertial density. The three-velocity is given by a simple average,

$$V_b^i dt = \frac{dt}{2} \left(\frac{U_i}{\sqrt{1 + U_i^2}} + \frac{U_{i-1}}{\sqrt{1 + U_{i-1}^2}} - V_g^i - V_g^{i+1} \right), \qquad (2.120)$$

while the zone-centered four-velocity and inertial mass are determined by Lagrange interpolation,

$$U_b^i = \frac{U_i dx_a^{i+1} + U_{i+1} dx_a^i}{dx_a^i + dx_a^{i+1}}, \tag{2.121}$$

$$\Delta M_{Ib}^i = \frac{\Delta M_I^i dx_a^{i+1} + \Delta M_I^{i+1} dx_b^i}{dx_a^i + dx_a^{i+1}}. \tag{2.122}$$

Next, in the same way that we sought an optimum density gradient before, we now wish to find an optimum derivative of the four-velocity $\nabla \tilde{U}^i$ at each zone. The maximum and minimum velocities around each zone are found:

$$U_{min} = \min\left\{U_{i-1}, U_i, U_{i+1}\right\}, \tag{2.123}$$

$$U_{max} = \max\left\{U_{i-1}, U_i, U_{i+1}\right\}. \tag{2.124}$$

We next define

$$\Delta U_{min} = \min\left(U_{max} - U_i, U_i - U_{min}\right), \tag{2.125}$$

$$\Delta U_b = U_b^i - U_b^{i-1}, \tag{2.126}$$

and select an optimum derivative,

$$\nabla \tilde{U}^i = \frac{\min\left\{\Delta U_{min}, |\Delta U_b|\right\}}{dx_a^i} \left(\frac{\Delta U_b}{|\Delta U_b|}\right). \tag{2.127}$$

The zone-centered four-velocity for momentum transfer $\tilde{U}^i$ is then evaluated,

$$\tilde{U}^i = \left(U_i + \frac{1}{2}\nabla\tilde{U}^i\left[dx_a^i + V_b^i dt\right]\right) \quad \text{if} \quad V_b^i > 0, \tag{2.128}$$

$$\tilde{U}^i = \left(U_i - \frac{1}{2}\nabla\tilde{U}^{i+1}\left[dx_a^i - V_b^i dt\right]\right) \quad \text{if} \quad V_b^i < 0. \tag{2.129}$$

The average transferred momentum $\Delta\Phi_S$ is then

$$\Delta\Phi_S^i = \Delta M_{Ib}^i \tilde{U}^i. \tag{2.130}$$

Finally, the updated momentum density is

$$S_i = S_i - (\Delta\Phi_S^i - \Delta\Phi_S^{i-1})/Vol_a^i. \tag{2.131}$$

2.4.17 Completion of the cycle

With the densities and momenta thus updated it is necessary once more to compute the pressures, densities and internal energies from the equation of state, followed by updated velocities U_i and V^i. The cycle is then completed and one is ready to advance to the next time step.

As demonstrated in the test problems to follow, the algorithm described in this subsection gives a reasonable description of the hydrodynamics even for strong shocks and velocities close to c. The principle shortcoming of the method given here is that the artificial viscous stress Q and the Lorentz factor W are not well enough centered in time to give full second-order accuracy. Some recent methods exist to center Q in time for Newtonian hydrodynamics. As discussed in the viscosity section above, we have not yet developed a method to center Q and W for fully relativistic flows.

2.5 Test problems

In view of the complexity of numerically evolving even Newtonian or special relativistic fluids, it is essential to have in hand a number of code calibrations and checks. These checks are the only way to judge the magnitude of the numerical errors inherent in finite difference schemes. Possible tests of particular interest in relativistic astrophysics have been summarized at various times (e.g. Hawley *et al.* [12, 13], Centrella and Wilson [4]). Here we summarize some of those tests which are most relevant to special or general relativistic problems.

2.5.1 Shocks and jump conditions

There is a simple class of relativistic solutions which involve one-dimensional shocks propagating through the grid. These are useful even as a test of multi-dimensional codes, particularly if one propagates the shock along various angles with respect to the grid. In the analytic solutions, the shock is a discontinuity separating regions of constant velocity and density V_1, ρ_1, ϵ_1 on one side from the post-shock regions with velocity V_2, ρ_2, ϵ_2 on the other side. Each region is separately isentropic, but the entropy is discontinuous across the shock, being higher in the post-shock region. The relations between the two states for the simple problems described here are derived by the application of conservation laws which are referred to as *jump conditions*. Here we introduce the following notation to denote conditions on either side of the shock,

$$\left[f\right] \equiv f_2 - f_1, \qquad (2.132)$$

where f_1 and f_2 are functions evaluated on the pre-shock and post-shock sides of the discontinuity, respectively. Thus, for a Newtonian shock front moving with constant velocity V_s we write [6, 12] the following jump conditions. The conservation of mass implies,

$$\left[\rho(V - V_s)\right] = 0. \tag{2.133}$$

The conservation of momentum flux gives,

$$\left[\rho(V - V_s)^2 + P\right] = 0, \tag{2.134}$$

and the conservation of energy condition leads to,

$$\left[\rho\left(\frac{V_s^2}{2} + \epsilon\right)(V - V_s) + PV\right] = 0. \tag{2.135}$$

From these relations the conditions on either side of the shock can be determined in terms of those of the other side.

The generalization to special relativity of these conditions requires the specification of a reference frame, and the application of the special relativistic equations previously derived. In a frame in which the shock is moving with a constant four-velocity U_s (and three-velocity $V_s = U_s/W_s$), the special relativistic generalization of the jump conditions can be written [25]

$$\left[D(VW_s - U_s)\right] = 0, \tag{2.136}$$

$$\left[(D + \Gamma E)U(VW_s - U_s) + (\Gamma - 1)EW_s/W\right] = 0, \tag{2.137}$$

$$\left[(D + \Gamma E)(UW_s - U_sW) + (\Gamma - 1)EU_s/W\right] = 0, \tag{2.138}$$

where we have specifically assumed an equation of state of the form of Eq. (2.18).

2.5.2 Relativistic wall shocks

The simplest test problems are those of an accelerating or stopping wall shock. These problems provide a good test of the hydrodynamic description of the jump conditions and shock heating in both Newtonian and special relativistic regimes. Wall shock problems simply involve the collision between a fluid of initial homogeneous density ρ_1 and a wall with reflecting boundary conditions at one end of the grid. A shock develops at the wall leaving behind a post-shock region consisting of a fluid at a density ρ_2 heated to a specific internal energy ϵ_2 as depicted in Figure 2.3.

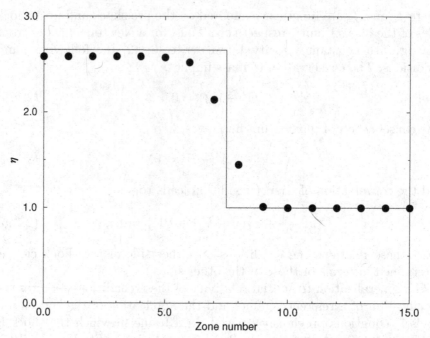

Fig. 2.3. Compaction $\eta = \rho_2/\rho_1$ from a relativistic stopping wall shock for a $\Gamma = 5/3$ fluid and initial conditions $U_1 = 1$, $D_1 = 1$, $\rho_1 = 1$, and $W = 1.8$.

2.5.3 Accelerating wall shocks

A more stringent test problem than the stopping wall shock for relativity codes is that of the accelerating wall shock. In this case a fluid initially at rest is impinged upon by a wall moving with relative velocity V_w. A shock then developes which accelerates away from the wall. Obviously, in the frame of the shock the two problems are invariant as to whether it is the wall or the fluid which is moving with respect to the grid. Nevertheless, in numerical simulations the problems are not the same. The accuracy of the numerical algorithm is not Lorentz invariant. This leads to an interesting test as discussed below.

The exact solution to the relativistic accelerating wall shock problem is given in [25]. For the shock jump conditions, with $U_1 = V_1 = 0$ and $W_1 = 1$ in the unshocked material and $U_2, V_2 = V_w, W_2$ denoting the post-heated material, one has,

$$\rho_1 V_s = \rho_2 W_2 (V_s - V_2), \tag{2.139}$$

$$\rho_2 h U_2^2 + P_2 = V_s \rho_2 h W_2 U_2 + P_1, \tag{2.140}$$

$$V_s(\rho_2 h W_2^2 - P_2) - \rho_2 h W_2 U_2 = V_s(\rho_1 h - P_1), \tag{2.141}$$

These lead to the following relations. The compression ratio, $\eta \equiv \rho_2/\rho_1$, becomes

$$\eta = \frac{P_2 - P_1 + \Gamma P_2 U_2^2/(\Gamma - 1)}{W_2(P_2 - P_1) - \rho_1 U_2^2}. \tag{2.142}$$

The velocity of the shock front in the frame of the unshocked fluid at rest is

$$V_s = \frac{P_2 - P_1 + \Gamma P_2 U_2^2/(\Gamma - 1)}{[\rho_1 + \Gamma W_2 P_2/(\Gamma - 1)]U_2}. \tag{2.143}$$

An auxiliary relation which can be used to find the post-shock pressure is

$$V_s\left(\rho_2 h W_2^2 - P_2 - \rho_1 - \frac{P_1}{\Gamma - 1}\right) - \rho_2 h W_2 U_2 = 0. \tag{2.144}$$

An iterative method for the solution of this set of equations is discussed in [25].

For numerical tests it is more convenient [5, 12, 13, 25] to consider the limit of zero pre-shock internal energy ($\epsilon_1 = P_1 = 0$). In this limit the jump conditions simplify to the following relations for the accelerated wall shock. The post-shocked pressure P_2 is

$$P_2 = \rho_1[\Gamma W_2^2 - (\Gamma - 1)W_2 - 1]. \tag{2.145}$$

The compression ratio is

$$\eta \equiv \frac{\rho_2}{\rho_1} = \left[\frac{\Gamma + 1}{\Gamma - 1} + \frac{\Gamma(W_2 - 1)}{\Gamma - 1}\right], \tag{2.146}$$

and the shock velocity is

$$V_s = \left(\frac{1 + \Gamma U_2^2/(\Gamma - 1)}{\rho_1 + \Gamma W_2 P_2/(\Gamma - 1)}\right)\frac{P_2}{U_2}. \tag{2.147}$$

Thus, once the pre-shock density ρ_1 and the velocity of post-shocked fluid $U_2 = W_2 V_2$ are specified, the remaining variables can be deduced. Note also, that in the laboratory frame, the shock front accelerates as the pressure increases behind the shock.

2.5.4 Accelerating shocks in the Eulerian gauge

Relativistic accelerating shocks are very difficult to describe accurately in the Eulerian gauge. Calculations are made of the coordinate density $D = W\rho$. Hence, the calculated coordinate compression ratio in this case is given by

$$\frac{D_2}{D_1} \equiv \eta_D = W_1\left[\frac{\Gamma + 1}{\Gamma - 1} + \frac{\Gamma(W_1 - 1)}{\Gamma - 1}\right], \tag{2.148}$$

$$= (W_1 + \Gamma W_1^2)/(\Gamma - 1). \tag{2.149}$$

Table 2.1 Summary of accelerating wall shock results

U	W	η_{true}	η_{calc}	Error (%)
0.001	1.0	4.00	3.97	0.8
1.0	1.414	7.1	7.2	1.1
2.0	2.24	16.	17.	7.2
4.0	3.87	43.3	~50.	15.
10.0	10.05	267.	~200.	30.

The coordinate compression ratio η is very large for large W_1. The zone-to-zone coordinate density ratio is also large in the shock. This leads to poor numerical accuracy. In the Eulerian representation $W \approx 2$ is about as high as can be tolerated (10% error). To calculate accelerating shocks the grid should be moved $(V_g = V_s)$ to keep the zoning better. For illustration, we have run accelerating shocks with a moving grid and obtain the following results shown in Table 2.1 for $\Gamma = 5/3$.

As U increases the results become more erratic, i.e. fluctuations in zone-to-zone density become large for high U. Hence, the mean compression ratio becomes poorly defined. Grid velocities should be chosen so as to avoid accelerating shocks with high velocities.

2.5.5 Stopping wall shocks

In the stopping wall shock, the fluid impinges on a stationary wall. This also has a simple solution [5, 12, 13]. The relation for the compression ratio is the same except that $W_2 \to W_1$,

$$\eta = \left[\frac{\Gamma + 1}{\Gamma - 1} + \frac{\Gamma(W_1 - 1)}{\Gamma - 1}\right]. \tag{2.150}$$

In this case the specific internal energy can be determined by the jump conditions or by equating the initial kinetic energy per nucleon to the internal energy per nucleon in the post-shock region. This becomes

$$\epsilon_2 = W_1 - 1, \tag{2.151}$$

from which the pressure in the post-shock region can be written analogously to the accelerating wall shock

$$P_2 = \rho_1[\Gamma W_1^2 - (\Gamma - 1)W_1 - 1]. \tag{2.152}$$

Table 2.2 Summary of stopping wall shock results

U	W	η_{true}	η_{calc}	Error (%)
0.001	1.0	4.00	4.02	0.5
0.458	1.1	3.864	3.865	0.03
1.0	1.41	3.57	3.53	1.1
1.732	2.00	3.25	3.20	1.5
10.0	10.	2.65	2.58	2.7
100.	100.	2.514	2.455	2.6

The main difference in this case is that the shock propagates with a constant velocity in the rest frame of the fixed wall,

$$V_s = V_1 \left/ \left(\frac{\rho_2}{\rho_1} \frac{1}{W_1} - 1 \right) \right. . \tag{2.153}$$

Though the form of these two problems is quite similar, in numerical calculations the accuracy of the method is not Lorentz invariant. In the Eulerian hydrodynamics discussed in this chapter, the calculated quantity in a numerical solution is D not ρ. Hence, the jump condition for stopping wall shocks becomes

$$\frac{D_2}{D_1} = \frac{\eta}{W_1}, \tag{2.154}$$

while for accelerating shocks in which the wall is in motion relative to the fluid we have

$$\frac{D_2}{D_1} = \eta W_2. \tag{2.155}$$

In practice, good results can be obtained for stopping wall shocks [16]. Figure 2.3 and Table 2.2 show examples of density calculated for a stopping wall shock. Even for for $W_1 = 100$ ($V_1 \approx 0.99995c$), the error in η is only a few percent. However, for accelerating shocks the accuracy falls below 1% even for $W_2 > 2$ as illustrated in Table 2.1.

2.5.6 Relativistic rarefaction

A second good test of the advection scheme is the relativistic rarefaction wave. This problem was first treated numerically by Centrella and Wilson [4] and later solved analytically by Thompson [25]. We summarize the solution of Thompson here.

In the rarefaction problem a high density region ρ_1 with internal energy ϵ_1 is separated by a thin membrane from a region of negligible density and internal energy, ρ_2, $\epsilon_2 = 0$. The membrane is removed at $t = 0$. A rarefaction wave then propagates into the high density region at the speed of sound c_s, as material freely expands into region 2. If region 2 is a true vacuum, the leading edge expands rapidly and cools to the point where the speed of sound vanishes $c_s = 0$.

The rarefaction solution begins with the definition of some auxiliary variables,

$$y \equiv \left(\frac{\Gamma}{(\Gamma - 1)}\frac{P}{\rho}\right)^{1/2} = y_1\left(\frac{\rho}{\rho_1}\right)^{(\Gamma-1)/2}. \tag{2.156}$$

We then have the density ρ and pressure P defined in the rarefaction in terms of the y,

$$\rho = \rho_1\left(\frac{y}{y_1}\right)^{2/(\Gamma-1)}, \tag{2.157}$$

and

$$P = P_1\left(\frac{\rho}{\rho_1}\right)^{\Gamma}, \tag{2.158}$$

where the quantity y can be solved in terms of U the spatial component of the four-velocity of the rarefaction wave,

$$y = \frac{f^2(y_1) - f^{\sqrt{\Gamma-1}}(U)}{2f(y_1)f^{(\sqrt{\Gamma-1}/2)}(U)}, \tag{2.159}$$

where the function f is defined,

$$f(x) \equiv x + (1 + x^2)^{1/2}. \tag{2.160}$$

This gives ρ and P as a function of U. The speed of sound in the rarefaction wave is also given by

$$c_s^2 = \frac{(\Gamma - 1)y^2}{1 + y^2}. \tag{2.161}$$

The velocity of material in the rarefaction wave can be given as a function of U

$$V = \frac{U}{W}, \qquad W^2 = (1 + U^2), \tag{2.162}$$

and the location of material at a given velocity is given by a similarity solution

$$\frac{x}{t} = V - \frac{c_s}{W(1 - c_s V)}. \tag{2.163}$$

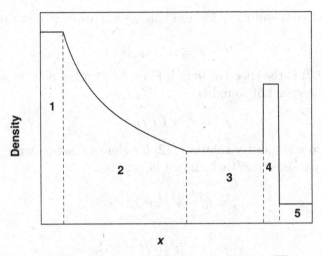

Fig. 2.4. Various regions in the shock tube problem. They are: (1) the undisturbed high density fluid; (2) the rarefaction wave; (3) a region of constant velocity and pressure which features a contact discontinuity separating regions of different density; (4) the shock itself; and (5) the undisturbed low density fluid.

2.5.7 Newtonian and relativistic shock tube

One of the most popular tests of numerical hydrodynamics is the one-dimensional Riemann shock tube. In this test, hot dense fluid on one side is separated by a membrane from cool rarefied gas on the other. The membrane is removed at $t = 0$ and the high density fluid pushes into the low density fluid, causing a rarefaction wave to move at the speed of sound back through the high density fluid. At the same time, the inflow of material into the low density region causes a shock to travel through the low density material. Thus, this test encorporates both the shock and rarefaction described in the previous problems.

Here, we first sketch the Newtonian shock tube solution. The generalization to relativity involves a straightforward application of the wall shock and rarefaction wave solutions given above and in [25].

As the fluid evolves, five distinct regions in the flow develop as depicted in the density solution on Figure 2.4. From high density to low density they are: (1) the undisturbed high density fluid; (2) the rarefaction wave; (3) a region of constant velocity and pressure which features a contact discontinuity separating regions of different density; (4) the shock itself; and (5) the undisturbed low density fluid. Thus, this problem tests both the advection scheme and the treatment of shocks.

The Newtonian solution for the rarefaction wave is described by the equation

$$x = (V - c_s)t \tag{2.164}$$

where $c_s(\rho, P)$ is the speed of sound. For a Newtonian polytropic gas with $P = k\rho^\Gamma$, the speed of sound is,

$$c_s^2 = \Gamma P / \rho. \tag{2.165}$$

This leads to a similarity solution [12] for the rarefaction wave as a function of the parameter x/t which can be written,

$$c_s = \eta^2 \frac{x}{t} + \left[1 - \eta^2\right](c_s)_0, \tag{2.166}$$

$$V = \left[1 - \eta^2\right][x/t - (c_s)_0], \tag{2.167}$$

where η is the Newtonian limit of the compression ratio ρ/ρ_1 from Eq. (2.142),

$$\eta = \left(\frac{\Gamma + 1}{\Gamma - 1}\right). \tag{2.168}$$

From these relations P and ρ can be determined from the polytropic equation of state and

$$\rho = \left[\frac{c_s^2}{\Gamma k}\right]^{1/(\Gamma-1)}. \tag{2.169}$$

The rarefaction wave connects with the intermediate contact discontinuity at the point at which the pressure and velocity of the wave equal that of the post-shock region. That is, there is a shock wave moving into the low density region with velocity

$$V = V_{low} + (P - P_{low})\left[\frac{(1 - \mu^2)}{\rho_{low}(P + \mu^2 P_{low})}\right]^{1/2}, \tag{2.170}$$

and there is a rarefaction wave moving into the high density region with a velocity

$$V = V_{high} - c_s^{high} + c_s. \tag{2.171}$$

The intersection of the two curves defined by these equations determines the velocity and pressure of the intermediate region.

Having found the velocity and pressure in the intermediate region, the Newtonian density in the intermediate region before the shock is given from the polytropic relation

$$\rho_m = \rho_{high}\left(\frac{P_{mid}}{P_{high}}\right)^{(1/\Gamma)}. \tag{2.172}$$

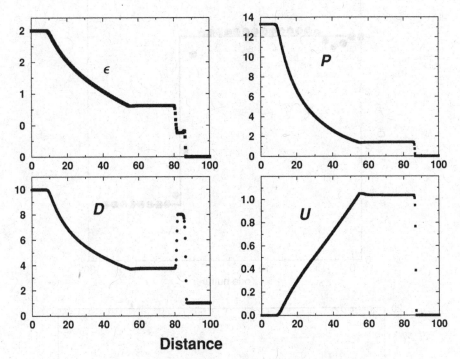

Distance

Fig. 2.5. Plots of specific internal energy ϵ, pressure P, density D, and four-velocity U for a $\Gamma = 5/3$ relativistic shock tube with $W_{max} = 1.43$.

However, the density in the post-shock region must be obtained by combining the jump conditions at the shock boundary to to obtain [12]

$$\frac{\rho_s}{\rho_{low}} = \left(\frac{\Gamma+1}{\Gamma-1}\frac{P_s}{P_{low}}+1\right)\Big/\left(\frac{\Gamma+1}{\Gamma-1}+\frac{P_s}{P_{low}}\right). \qquad (2.173)$$

In all of the above, the internal energy ϵ can be obtained from the pressure and density

$$\epsilon = P/[(\Gamma-1)\rho]. \qquad (2.174)$$

The relativistic generalization of the shock tube problem involves a straightforward application of the relativistic rarefaction and wall shock problems as sketched above and in [25]. Figure 2.5 shows results from the one-dimensional code described in this chapter for a $\Gamma = 5/3$ gas with 500 spatial zones. The code was run until $t = 50$ and the maximum Lorentz factor for this test is $W_{max} = 1.43$. The agreement with the exact solution is generally better than 1%.

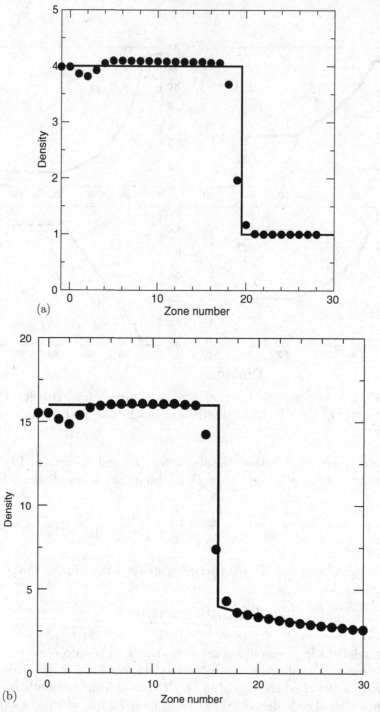

Fig. 2.6. Density versus zone number for a nonrelativistic Noh collapse in (a) one, (b) two, and (c) three dimensions with initial conditions $V_0 = -1$, $\rho_0 = 1$.

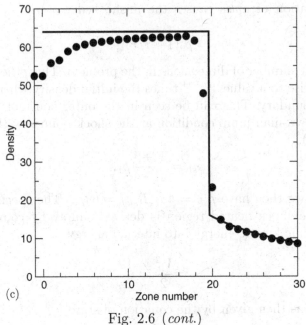

(c)

Fig. 2.6 (*cont.*)

2.5.8 Newtonian Noh solution

The cylindrical and spherical versions of the Newtonian wall shock problem are referred to as the Noh shocks. The planar version is just the Newtonian limit of the relativistic wall shock discussed in Section 2.5.5. Although this set of problems is nonrelativistic, we mention it here as it is a good test for the shock capturing (artificial viscosity) scheme. It is also a good test for spherical supernova collapse codes to be discussed in Chapter 5.

In the Noh problem an initially uniform gas of zero internal energy moves toward the origin with constant radial velocity. As the collapse ensues, a shock starts in the center as the infalling matter compresses and heats to large pressure. Many codes designed to do spherical collapse do very poorly on this problem. The reason is that simple artificial viscosity treatments tend to produce large heating at the very center resulting in a region of unphysical high heating and low density. Hence, this is an excellent problem in which to test codes designed to do spherical collapse.

The Noh problem, as depicted in Figure 2.6, begins at $t = 0$ with a $\Gamma = 5/3$ gas of uniform density $\rho_0 = 1$ collapsing with a velocity $V_0 = -1$. The accreting material produces an outward going shock with velocity $V_{sh} = 1/3$ for this equation of state from the Newtonian ($W = 1$) limit of Eq. (2.153). The shock radius is therefore located at $R_{sh} = tV_{sh} = t/3$.

The conservation of mass causes the unshocked material to develop a radial profile

$$\rho_1 = \rho_0(1 + t/r)^{(n-1)}, \tag{2.175}$$

where n is the number of dimensions in the problem. Thus, the unshocked density will rise to a value $4^{(n-1)}$ times the initial density when it reaches the shock boundary. This can be seen in the outer zones of Figure 2.6. From the Newtonian jump condition at the shock boundary (Eq. (2.150) with $W \to 1$),

$$\rho_{sh} = \frac{\Gamma + 1}{\Gamma - 1}\rho_1, \tag{2.176}$$

for $\Gamma = 5/3$ we then have $\rho_{sh} = 4\rho_1(R_{sh}) = 64\rho_0$. The specific internal energy in the shock heated region is derived simply by converting the infalling specific kinetic energy into internal energy,

$$\epsilon = \frac{V_0^2}{2} = \frac{1}{2}. \tag{2.177}$$

The pressure is then given by the equation of state,

$$P_{sh} = (\Gamma - 1)\rho\epsilon, \tag{2.178}$$

which in the present example is $64/3$.

Some results from numerical calculations in one, two, and three dimensions are shown in Figure 2.6. These can be summarized as follows. For the first several zones, the artificial viscosity overheats the zones. Pressure equilibrium then implies that the average densities of the first few zones are below the values of 4, 16, and 64 expected by the jump condition (2.176) for this example. (In this example we have 3.88, 15.03, and 50.9.) However, the densities at several zones behind the shock are 4.022, 16.06, and 63.7, in good agreement with expectations.

These results emphasize two points: (1) there is a need to suppress preheating in smooth inward compressing flow; (2) it is necessary to resolve shocks sharply so that the shock does not become confused with the $\partial\rho/\partial r$ gradient outside the shock.

2.5.9 Homologous Newtonian collapse

Another related useful problem is that of homologous collapse and uniform compression. In this case a uniform density distribution of radius r_0 is taken to have an the initial collapse velocity of

$$V = -V_0(r/r_0). \tag{2.179}$$

The collapse should be self similar. A fluid with uniform initial density ρ_0 should remain uniform but grow exponentially,

$$\rho = \rho_0 e^{\{3V \cdot t/r \cdot\}}. \tag{2.180}$$

In particular, there should be no heating.

2.5.10 Other test problems

Another good test of a code is how many zones are required to resolve a static star. We tested the accuracy of the three-dimensional code described in Chapter 3 and used in Chapter 7 by calculating static neutron star properties. For example, we compared the central density calculated by a highly zoned spherical computer program with the three-dimensional code using significantly fewer zones. Even with only 14 zones in radius the error in central density was less than 0.1%. The error increased to ~1.0%, however, for a calculation with eight zones in the radial direction.

For codes written in spherical coordinates, tests should be made of both planar shocks and linear slabs passing through the origin. Neither the shock nor the slab should become distorted by passing through the origin in spherical coordinates. For a discussion of advection in curvilinear coordinates see Section 3.4.3.

We conclude this section with Table 2.3 which shows some other possible test problems and why they are useful. For more explanation see [5]. It is strongly recommended that one apply all of the even marginally relevant tests to a code before applying it to a real problem. This is usually the only way to quantify the limits and accuracy of a numerical code.

2.6 Application to heavy ion collisions

We wish to conclude this chapter with an example of a real application of special relativistic hydrodynamics. Most of the numerical calculations described in this book are in the context of astrophysical applications. There are several environments of astrophysical phenomena which can be adequately modeled using special (as opposed to general) relativistic hydrodynamics. Some examples include the propagation of relativistic jets in the interstellar medium, shock waves in accretion disks far from a neutron star, hydrodynamic processes in radiation dominated stars, collisions of high velocity stars and/or white dwarfs, and the early stages of core-collapse supernova before the formation of the compact neutron star (or far from the neutron star at late times). There is one application, however, which occurs in the laboratory, and therefore is subject to closer scrutiny. It is also an important means to study terrestrially the equation

Table 2.3   Summary of useful test problems

Test	Points tested
One dimensional	
Standing wall shock [12]	artificial viscosity
Accelerating wall shock [12]	artificial viscosity/Lorentz invariance
Relativistic rarefaction [25]	advection
Riemann shock tube [25]	advection and artificial viscosity
Noh collapse [17]	artificial viscosity
Homologous collapse [12]	advection
Uniform slab advection [12]	advection of D and S_x
Two and three dimensional	
Spherical pressureless slab $V^r = 0$	stays spherical
Linear slab in spherical mesh	curvilinear advection
Noh collapse [17]	artificial viscosity
Wall shocks at 45° to mesh	Cartesian advection
Wall shocks in spherical mesh	spherical advection at $r = 0$
Taylor instabilities	overall hydrodynamics
Helmholtz instabilities	overall hydrodynamics
Orbiting equilibrium tori [12, 13]	pressure and gravitational forces
Jeans ellipsoids [9]	Newtonian hydrodynamics
Newtonian Homogen. Ellipsoids [21]	density and velocity profiles
Equilibrium of Gen. Rel. Fluids [22]	stability
Pulsation frequencies of relativistic neutron stars [10]	dynamics near equilibrium
General relativistic spherical dust collapse [18, 23]	general relativistic dynamics
Nonspherical dust infall to a black hole [19]	general relativistic dynamics

of state relevant to core-collapse supernovae and neutron stars. This is the relativistic collision of two heavy nuclei.

Experiments have been carried out, for example at the Berkeley Bevalac, in which heavy ions, e.g. ^{139}La, ^{197}Au, were made to collide at laboratory energies of 0.5 to 1.3 GeV/nucleon and more. The associated barycenter energies and nuclear temperatures are roughly comparable to the binding energy per nucleon and central temperature of the nascent neutron star formed during a supernova collapse. Hence, a hydrodynamic description of such heavy ion collisions is expected to be useful as a means to constrain the equation of state to be used used to model supernova collapse.

However, in order to make such comparisons, it is important to have some confidence that colliding heavy ions can indeed be modeled as hydrodynamic Fermi nuclear fluids plus a possible bosonic (e.g. pion) contribution. This is not immediately obvious. For example, the hydrodynamic equations described herein implicitly assume a short particle mean free path so that the particle distribution functions can be replaced by their moments. One consequence of this is that fluids do not mix during a hydrodynamic single-fluid collision. However, the nuclear shell model indicates that nucleons can be viewed as independent particles in a mean field. One also knows that quantum mechanically the nuclear wave function must be viewed as a Slater determinant of indistinguishable particles which can easily mix during the collision. This has led to confusion as to the applicability of hydrodynamics to simulate heavy ion nuclear collisions.

We wish to emphasize that this apparent dichotomy does not preclude a hydrodynamic description for heavy ion collisions. The existence of a mean nuclear field is itself the result of many nucleon-nucleon interactions. Indeed, the experimental evidence discussed below is consistent with multiple nucleon–nucleon interactions during heavy ion collisions. We show below that this is further enhanced by interactions with the hadronic excitations during the collision.

Although a single-fluid hydrodynamic model does assume a short mean free path, this assumption can be avoided by utilizing multiple fluids and/or viscosity/heat flow to describe the various interactions involved during the collisions. In this way, depending upon the coupling between the colliding fluids, one can go from the limits of "transparent" nuclei to complete nuclear stopping. Even if the hydrodynamic equations become inappropriate for some component of the fluid (e.g. pions or neutrinos), one can always evolve the full Boltzmann equation for the distribution functions. Indeed, as we shall see in the chapter on supernovae, this becomes a necessity for the neutrino fluid as it escapes from the proto neutron star.

For application to heavy ion collisions it is straightforward enough to consider the nuclei as colliding Fermi nuclear fluids. The only aspects missing from a classical hydrodynamic treatment are those arising from explicit quantum mechanical effects such as asymmetrization and tunneling. Nevertheless, the bulk classical properties of the collision can be represented by classical hydrodynamics. Indeed, the model described below is in excellent agreement with the distribution of pions from relativistic heavy ion collisions. The collision of two nuclei inevitably exhibits a high degree of stopping as the two nuclear fluids merge. One wishes to know whether this picture is justified on the basis of the microscopic exchange of energy and momentum in the nucleon–nucleon interactions during a heavy ion collision. This can be described in the context of a three-fluid hydrodynamic model [31] which we now describe. This model explains the stopping of central heavy ion collisions and thus justifies the application of hydrodynamics to the process. This makes heavy ion collisions a viable experimental probe of the supernova equation of state.

2.6.1 Hydrodynamics and heavy ion stopping

Since the one-fluid hydrodynamic model [16] necessarily implies complete stopping of the head-on collisions of incoming nuclei, Wilson (1996) investigated that assumption by the introduction of a three-fluid, one-dimensional (planar) hydrodynamic model to study partial nucleon stopping. In this work it was demonstrated that physically reasonable nucleon–nucleon energy exchange could account for the high degree of stopping observed in heavy ion collisions and justify the good agreement obtained in one-fluid simulations.

In this model two of the fluids represent the initially right- and left-moving nucleonic matter in the barycenter of mass. In proton–proton collisions a small fraction of the proton kinetic energy appears as pions of low kinetic energy in the barycenter frame. Hence, the third fluid is associated with the energy produced by collisions producing pions, kaons, and other hadronic matter. The equations of motion for this system are thus,

$$\frac{\partial D_i}{\partial t} + \frac{\partial}{\partial z}(D_i V_i) = \dot{D}_i^c, \tag{2.181}$$

$$\frac{\partial E_i}{\partial t} + \frac{\partial}{\partial z}(E_i V_i) + (\Gamma - 1)\frac{E_i}{W_i}\left(\frac{\partial}{\partial z}V_i + \frac{\partial W_i}{\partial t}\right) = \dot{E}_i^c, \tag{2.182}$$

$$\frac{\partial S_i}{\partial t} + \frac{\partial}{\partial z}(S_i V_i) + \frac{\partial}{\partial z}\left((\Gamma - 1)\frac{E_i}{W_i}\right) = \dot{S}_i^c, \tag{2.183}$$

where $i = r, l, e$ represents the rightward-moving, leftward-moving and energetic collision-produced fluids, respectively. The motion is taken along the z direction, so the four-velocity is

$$U^\alpha = W(1, 0, 0, V). \tag{2.184}$$

Also, as usual, the fluids are taken to be perfect gases so that the pressure can be written locally as

$$P = (\Gamma - 1)\frac{E}{W}. \tag{2.185}$$

In the above, the collision terms are derived by assuming a simple form that conserves energy and momentum,

$$\dot{E}_r^c = \dot{E}_l^c = 0, \tag{2.186}$$

$$\dot{E}_e^c = \dot{E}_{lr}^c + \dot{E}_{re}^c + \dot{E}_{le}^c, \tag{2.187}$$

where

$$\dot{E}_{lr}^c = \sigma_0 \delta E_0 D_l D_r |\frac{V_r - V_l}{1 - V_r V_l}|, \tag{2.188}$$

$$\dot{E}_{re}^c = \sigma_e \delta E_e D_r (E_e/M_n)c, \tag{2.189}$$

$$\dot{E}_{le}^c = \sigma_e \delta E_e D_l (E_e/M_n)c. \tag{2.190}$$

For collisions of right- and left-moving fluids the following changes in momentum density are consistent with the energy exchange,

$$\dot{S}_r^c = -\dot{S}_r^c = -\dot{E}_{lr}^c \left[\frac{\partial T}{\partial E_e} / \left(-\frac{\partial T}{\partial S_r} + \frac{\partial T}{\partial S_l} \right) \right], \tag{2.191}$$

$$\dot{S}_e^c = 0, \tag{2.192}$$

where the time–time component of the energy momentum tensor is written

$$T \equiv \sum_i T_i = \sum \left[(D_i + \Gamma E_i) W_i + (\Gamma - 1) \frac{E_i}{W_i} \right]. \tag{2.193}$$

For the interaction of the right-moving fluid with the energy fluid we have,

$$\dot{S}_r^c = -\dot{S}_e^c = -\dot{E}_{re}^c \frac{\partial T}{\partial E_e} \left[\frac{K_r}{K_r - I_r} / \left(-\frac{\partial T}{\partial S_r} + \frac{\partial T}{\partial S_e} \right) \right.$$
$$\left. - \frac{I_r}{K_r + I_r} \left\{ S_r / \left(\left(\frac{\partial T}{\partial D_r} \right) (D_r + \Gamma E_r) \right) \right\} \right], \tag{2.194}$$

$$\dot{D}_r^c = -\dot{D}_e^c = -\dot{E}_{re}^c \frac{\partial T}{\partial E_e} \left(\frac{I_r}{K_r + I_r} / \frac{\partial T}{\partial D_r} \right), \tag{2.195}$$

where K_r is kinetic energy and $I_r = T_r - K_r = D_r + E_r$ is the non-kinetic energy of the right-moving fluid. A similar equation holds for the left-moving fluid.

The nucleon interaction cross-section σ_0 and the energy exchange δE_0 are determined from proton–proton data [31] to be

$$\sigma_0 = 2.4(\bar{U} - 1)^{1/10} \ \text{fm}^2, \tag{2.196}$$

$$\delta E_0 = -3.0 + 1.61 \log (6.5 + \bar{U}/2) \ \text{GeV}, \tag{2.197}$$

where $\bar{U} \equiv |U_r - U_l|$. At CERN energies (200 GeV/nucleon), for example, $\delta E_0^{max} = 1.4$ GeV and $\delta E_0^{mean} = 0.8$ GeV. At RHIC energies $\delta E_0^{max} = 4.2$ GeV and $\delta E_0^{mean} = 3.0$ GeV.

The corresponding cross-section and exchange energy is neither well known nor well defined for the nucleon/energy-fluid interaction. An order of magnitude estimate is $\sigma_e = 4.0$ fm^2, and $\delta E_e = 1.5$ GeV. Fortunately, results are rather insensitive to the product $\sigma_e \delta E_e$ over a broad range of values.

The equations above for $\dot{D}, \dot{E}$, and $\dot{S}$ are solved by the same methods as described in Section 2.4.1. The energy, density, and momentum exchanges are small per time cycle. Hence, they can be evaluated after the advection step (5). By use of the exchange terms, the heat flow and viscosity are accounted for at a microscopic level.

Calculations performed in one dimension as outlined above should represent the on-axis behavior of central collosions very well for the short time over which the calculations are made. Nuclei are initially represented by the usual nuclear profiles from electron scattering. The initial rest mass distribution is

$$\rho = \frac{0.16 \text{ N/fm}^3}{1 + \exp[(z - z_0)/\delta]}, \qquad (2.198)$$

with $z_0 = 5.2$ fm and $\delta = 0.56$ fm. For CERN collisions in the center-of-rapidity frame (used here) the initial Lorentz factor is $W = 10.41$ which corresponds to 200 GeV/nucleon in the laboratory frame.

In [31] an investigation was made into the nonlinear effects of scattering from the hadronic fluid by performing calculations with σ_e set to either zero or a finite value. Calculations with $\sigma_e = 0$ represent collisions in which only the nucleon–nucleon interactions occur, i.e. there are no additional interactions with the generated hadronic excitations. At late times, when the nuclei have passed through each other, the energy fluid density has its peak value. However, in the calculation with with $\sigma_e = 0$ the nucleonic fluid distributions were almost unaltered. The Lorentz factors of the nucleonic fluid only decreased from 10.4 to 9, while the density of the nucleonic fluid has increased by about a ratio of 10.4/9. Only about 20% of the initial kinetic energy was given to the energy fluid and the velocity of the energy fluid is small just after formation. Earlier in the collision there was some inward motion of the energy fluid induced by the nucleons imparting inward momentum.

Figure 2.7, however, shows the energy fluid density distribution for a calculation with $\sigma_e = 4.0$ fm^2 at the time at which the density is a maximum. In this case, over 99% of the baryonic initial kinetic energy is transferred to the energy fluid. Also shown in Figure 2.7 is the density

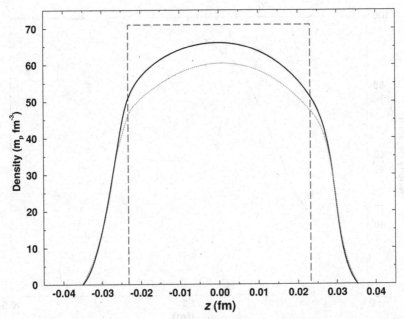

Fig. 2.7. Densities versus position at $t = 1$ fm/c for a three-fluid collision at 200 GeV/nucleon with $\sigma_e = 4.0$ fm^2. The solid curve is the total mass energy density of the energy fluid ($D_e + E_e$) while the dotted curve is the internal energy alone E_e. The dashed curve shows the mass energy density that would occur in a one-fluid model at the same time.

profile that would result from a one-fluid model for colliding uniform-density matter which initially had the same mean density as the nuclei had in this three-fluid model. The density distribution in the the three-fluid model has nearly the same peak but is slightly more diffuse in space than the one-fluid model. This justifies the validity of the one-fluid model.

Since the cross-section for the nucleon/energy-fluid interaction is somewhat *ad hoc*, calculations were made for a variety of σ_e values in [31]. The fraction of kinetic energy lost from the nucleons is shown in Figure 2.8. It is seen that the stopping efficiency is rather insensitive to σ_e.

Given that the predicted stopping is insensitive to the value of σ_e, calculations were also extended to energies characteristic of RHIC, i.e. initial Lorentz factor ~100. For $\sigma_e = 0$, only 7.4% of the nucleon energy is converted to collisional energy (the energy fluid). The results with $\sigma_e = 4.0$ fm^2 are presented in Figure 2.9. The stopping at the RHIC energy is similar to the stopping at CERN energies. Of the initial kinetic energy, 98% is converted into the energy fluid. The fraction of kinetic energy deposited in the energy fluid for the RHIC-like calculations is insensitive

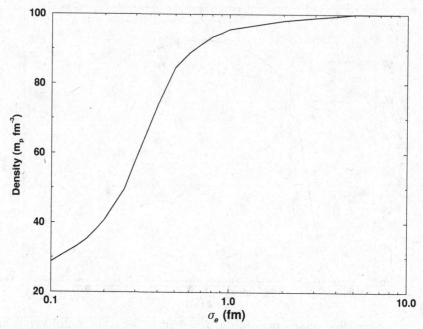

Fig. 2.8. Fraction of initial kinetic energy converted into the energy fluid as a function of σ_e for collisions with 200 GeV/nucleon.

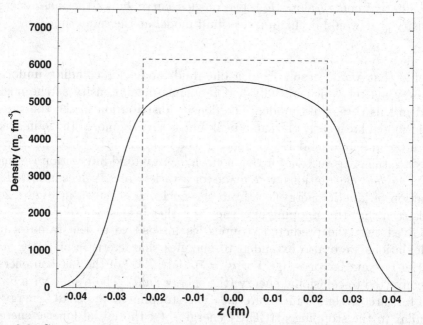

Fig. 2.9. Same as Figure 2.7 but for three-fluid collisions at RHIC-like energies (Lorentz factor 100).

to σ_e, similar to the results in Figure 2.7. For the case of $\sigma_e = 0$, at both CERN and RHIC energies, the depositied energy density varies almost linearly with the magnitude of δE_0. For the nominal value of $\sigma_e = 4.0$ fm^2, the deposited energy in both cases is insensitive to δE_0 (e.g. a reduction of δE_0 by a factor of two produces negligible changes in the total energy exchange).

2.6.2 Nuclear fluid plus pions

Having justified the high stopping power implied in one-fluid simulations, we now consider one-fluid nuclear collisions as treated by McAbee and Wilson [16]. In those simulations the baryonic component was described with relativistic Navier–Stokes equations ((2.32)–(2.34)) in three spatial dimensions as outlined in Section 2.2.2. The formation and evolution of pions was computed in the context of Landau–Migdal theory to determine the effective energies and momenta of the pions. The mean field model is based upon the work of Friedman, Pandharipande and Usami [8], whereby the pion energy is given by the following dispersion relation

$$\epsilon^2 = m_\pi^2 + p^2 \left(1 + \frac{\Lambda^2 \chi}{1 - g' \Lambda^2 \chi}\right), \tag{2.199}$$

where

$$\Lambda^2 \chi = -\frac{4.52 \omega \rho}{m_\pi (\omega^2 - \epsilon^2)} \exp\left\{-2\left(\frac{p}{7 m_\pi}\right)^2\right\}, \tag{2.200}$$

and

$$\omega = \sqrt{m_\Delta^2 + p^2} - m_N. \tag{2.201}$$

In this equation, M_π, m_Δ and m_N are the pion, delta, and nucleon rest masses, respectively.

An important quantity in the above is the Landau parameter g'. This is a density dependent quantity, taken to vary linearly with $\eta \equiv (\rho/\rho_0)$,

$$g'(\eta) = g_1 + g_2 \eta. \tag{2.202}$$

A key aspect of hydrodynamic simulations of the heavy-ion data is the determination of this quantity from fits to pion data at low impact parameter. Once g_1 and g_2 are fixed, the effect of pions on the equation of state can be determined and the influence of the pions on supernova evolution can be analyzed.

A comparison with experimental data requires specification of the pion distribution function. The evolution of the distribution function f for the pions is described by,

$$\dot{f} + v^\mu \frac{\partial f}{\partial x^\mu} + F^\mu \frac{\partial f}{\partial p^\mu} = \frac{v\rho\sigma_{sc}}{\gamma} \left(\frac{1}{4\pi} \int f(p')dp' - f \right) + \frac{v\rho\sigma_{abs}}{\gamma}(f_{BE} - f),$$

(2.203)

where f_{BE} is the equilibrium Bose–Einstein distribution. The pion three-velocity in the local matter rest frame is given as

$$v^i = \left(\frac{\partial \epsilon}{\partial p} \right) \frac{p^i}{p},$$

(2.204)

and the force exerted by the mean field is

$$F_\mu = -\left(\frac{\partial \epsilon}{\partial \rho} \right) \frac{\partial \rho}{\partial x^\mu}.$$

(2.205)

The isospin-averaged cross-sections for pion absorption and scattering were determined [16] from a fit to experimental cross-sections,

$$\sigma_{abs} = \frac{1.75 C(\eta)}{1 + 6(\sqrt{p^2 + m_\pi^2} - m_\pi)} \text{ fm}^2,$$

(2.206)

where $\eta \equiv \rho/\rho_0$ with $\rho_0 = 0.16 \text{ fm}^{-3}$, and

$$C(\eta) = \frac{(2/3)\eta}{1 + (\eta/3)} + \frac{(1/3)\eta^3}{1 + (\eta^4/16)},$$

(2.207)

and

$$\sigma_{sc} = \frac{\sigma_0}{1 + \eta} + \frac{8\eta}{(1 + \eta)(1 + p/p_c)} \text{ fm}^2,$$

(2.208)

where

$$\sigma_0 = \frac{40(p/p_a)^5}{1 + (p/p_a)^{10}} + \frac{8(p/p_b)^5}{1 + (p/p_b)^{10}}.$$

(2.209)

The following parameters were found to fit the data, $p_a = 290$ MeV/c, $p_b = 1453$ MeV/c, $p_c = 931.5$ MeV/c.

To solve for the distribution function, a Monte-Carlo method was applied in [16] to the Boltzmann equation.

2.6.3 Solving the Navier–Stokes equation with pions

Once the pion distribution function is specified, the hydrodynamic evolution can be obtained. To solve the viscosity and heat flow parts of the hydrodynamic equations, a Lorentz boost is made at each zone to the fluid frame of the state variables. Then the stress heat flow tensor Σ^μ_ν is

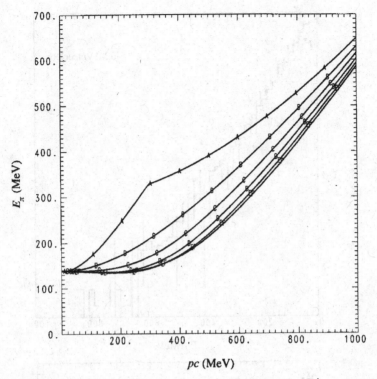

Fig. 2.10. Energy momentum dispersion relations for $g' = 0.5 + 0.06\eta$ (Eq. (2.199)) and with the density parameter η ranging from 0 to 5 in steps of one [16]. (Used by permission of *Nuclear Physics A.*)

evaluated (including the flux limiter corrections for the baryonic part). The pion fluid frame distribution function $f(p, x)$ is also used to form the pion stress tensor,

$$\Sigma^{\mu}{}_{\nu}(\text{pions}) = \frac{1}{2} \int f(p) g_{\mu\epsilon}(v^{\epsilon} p^{\mu} + v^{\nu} p^{\epsilon}) d^2 p. \qquad (2.210)$$

The total viscous stress is then boosted back to the computational frame and used to solve the Navier–Stokes equations described in Section 2.2.2.

The pion dispersion relation was solved for various densities relevant to the collisions. Some representative results are presented in Figure 2.10. The coupling constant in the pion dispersion relation was varied to find a good fit to the pion experimental data. The data were best fit with a value of $g' = 0.5 + 0.06\eta$.

The final results are summarized in Figure 2.11 and Table 2.4. The pion spectra summarized in Figure 2.11 agree well with the observations when the pion coupling is properly chosen.

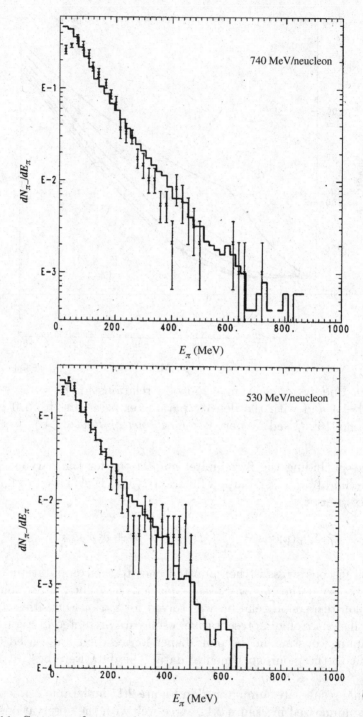

Fig. 2.11. Spectra of π^- in the center of mass compared to experimental data in the range 60° to 120°. (Used by permission of *Nuclear Physics A*.)

Table 2.4 Experimental and calculated π^- multiplicities for several collision energies from [16]

Energy (GeV/nucleon)	N_{π^-}		Average KE$_\pi$ (MeV)	T$_{peak}$ (MeV)	η_{peak}
	Experimental	Calculated			
0.246	–	0.56	64	27	2.3
0.520	4.0	4.00	77	47	3.0
0.740	8.5	8.32	87	57	3.4
1.350	21.3	21.23	106	75	3.5

Having determined the Landau coupling parameter $g'(\eta)$ in this way, the supernova equation of state was modified to incorporate pions in the same way. This change was found to enhance greatly the supernova explosion, illustrating the importance of heavy-ion collisions as a check on the dynamics of fluids in relativistic environments.

References

[1] P. Anninos, private communication (2002).

[2] D. S. Balsara, *J. Comput. Phys.*, **114**, 284 (1994).

[3] D. S. Balsara, *J. Comput. Phys.*, **174**, 614 (2001).

[4] J. M. Centrella and J. R. Wilson, *Astrophys. J.*, **273**, 428 (1983); *Astrophys. J. Suppl. Ser.*, **54**, 229 (1984).

[5] J. M. Centrella, S. L. Shapiro, C. R. Evans, J. F. Hawley and S. A. Teukolsky, in *Dynamical Spacetimes and Numerical Relativity*, ed. J. M. Centrella (Cambridge University Press, New York, 1986), pp. 328–344.

[6] R. Courant and K. O. Friedrichs, *Supersonic Flow and Shock Waves*, (Springer, New York, 1976).

[7] C. Eckart, *Phys. Rev.*, **58**, 919 (1940).

[8] B. Friedman, V. R. Pandharipande and Q. N. Usami, *Nucl. Phys. A*, **372**, 483 (1981).

[9] D. L. Gilden and S. L. Shapiro, *Astrophys. J.*, **287**, 728 (1984). *Astrophys. J. Suppl. Ser.*, **55**, 211 (1984).

[10] E. N. Glass and L. Lindblom, *Astrophys. J. Suppl. Ser.*, **53**, 73 (1984).

[11] A. Harten, B. Engquist, S. Osher and S. Chakrabarty, *J. Comput. Phys.*, **71**, 231 (1987).

[12] J. F. Hawley, L. L. Smarr and J. R. Wilson, *Astrophys. J.*, **277**, 296 (1984).

[13] J. F. Hawley, L. L. Smarr and J. R. Wilson, *Astrophys. J. Suppl. Ser.*, **55**, 211 (1984).

[14] B. I. Jun, private communication, 2002.

[15] J. M. Marti and E. Müller, *Living Rev. Relativ.*, **2**, (1999), http://www.livingreviews.org/Articles/Volume2.

[16] T. L. McAbee and J. R. Wilson, *Nucl. Phys. A*, **576**, 626 (1994).

[17] W. F. Noh, *J. Comput. Phys.*, **72**, 78 (1987).

[18] L. Petrich, S. L. Shapiro and S. A. Teukolsky, *Phys. Rev.*, **D31**, 2459 (1985); **D33**, 2100 (1986).

[19] L. Petrich, S. L. Shapiro and I. Wasserman, *Astrophys. J. Suppl. Ser.*, **58**, 297 (1985).

[20] P. L. Roe, *J. Comput. Phys.*, **43**, 537 (1981).

[21] R. A. Saenz and S. L. Shapiro, *Astrophys. J.*, **244**, 1033 (1981); **298**, 58 (1985).

[22] S. L. Shapiro and S. A. Teukolsky, *Astrophys. J.*, **234**, L177 (1979); **235**, 199 (1980).

[23] S. L. Shapiro and S. A. Teukolsky, *Astrophys. J.*, **298**, 34 (1985); **298**, 58 (1985).

[24] L. Smarr, C. Taubes and J. R. Wilson, in *Essays in Relativity: A Festschrift for Abraham Taub*, ed. F. Tipler (Academic Press, New York, 1980), pp. 157–183.

[25] K. W. Thompson, *J. Fluid Mech.*, **171**, 365 (1986).

[26] B. Van Leer, *J. Comput. Phys.*, **32**, 101 (1979).

[27] J. Von Neumann and R. D. Richtmer, *J. Appl. Phys.*, **21**, 232 (1950).

[28] S. Weinberg, *Gravitation and Cosmology* (Wiley, New York, 1973).

[29] J. R. Wilson, *Astrophys. J.*, **163**, 209 (1971).

[30] J. R. Wilson, in *Sources of Gravitational Radiation*, ed. L. Smarr (Cambridge University Press, Cambridge, 1979), p. 423.

[31] J. R. Wilson, *Nucl. Phys. A*, **606**, 339 (1996); **606**, 1317 (1996).

3

General relativistic hydrodynamics

3.1 General relativity

As in special relativity, there are two basic coordinate choices which have been employed to describe hydrodynamic flow in curved spacetime [2]. One choice is to force the spatial coordinate trajectories to coincide with the matter flow. This is the Lagrangian or comoving gauge. Until Wilson (1972) [4], all relativistic hydrodynamics employed this gauge (cf. [3] for a review). The other choice is to remove the comoving restriction and evolve the full Eulerian hydrodynamics equations. By using the gauge freedom in the choice of the ADM metric shift vector β^i (and/or the coordinate grid velocity V_g^i) it is possible [1, 2] to recover partially the comoving limit even in the Eulerian gauge.

The Lagrangian approach has some advantages, particularly for problems with spherical symmetry. For one, the entropy follows the mass points and is only changed by shocks or an external heat source (i.e. heat flow and viscosity). Another advantage is that velocity and density are related in a simple fashion. We will discuss an application of the Lagrangian method in Chapter 5 on supernova models.

There are, however, some significant disadvantages to the Lagrangian coordinates. Shock-wave discontinuities lead to metric derivative discontinuities, and vorticity leads to a nonlocal connection of fluid elements. Hence, comoving coordinate lines can become tangled and no longer simply related to spatial coordinates. For example, see the section in Chapter 2 on accelerating shock waves.

The advantage of an Eulerian description is that the metric can remain smooth and well defined even in the presence of complicated fluid flow. Hence, it is the best suited gauge for multi-dimensional flow. Its

75

disadvantage, however, is that steep gradients in state variables can be difficult to resolve numerically.

In practice we recommend only to use Lagrangian comoving coordinates for situations of such symmetry that only one spatial coordinate is necessary. The steep gradients in Eulerian systems can be managed by moving the coordinates so as to follow the matter as closely as possible, either by a judicious choice of the ADM shift vector or by employing a mixed Eulerian–Lagrange scheme as described below.

As noted in Chapter 1, we will make use of geometrized units ($G = c = 1$). We also make extensive use of the metric in ADM $(3 + 1)$ form,

$$ds^2 = -(\alpha^2 - \beta_i \beta^i)dt^2 + 2\beta_i dx^i dt + \gamma_{ij} dx^i dx^j, \tag{3.1}$$

where we take Greek indices to run over four coordinates and Latin indices to run over spatial coordinates. α is called the lapse function, β_i is the shift vector, and γ_{ij} is the three-space metric. We define

$$\gamma = \sqrt{\det(\gamma_{ij})}. \tag{3.2}$$

We also make use of the general relation for ADM coordinates,

$$\det(g_{\alpha\beta}) = -\alpha^2 \gamma^2, \tag{3.3}$$

in presenting the equations of hydrodynamics.

3.2 General relativistic hydrodynamics

Here we present the equations of hydrodynamics in Eulerian form. In the coming chapters, the only place we will use a Lagrangian coordinate system is in the chapter on spherical supernova collapse. There, we will develop the appropriate equations.

3.2.1 State variables

The basic hydrodynamic state variables we define as follows.

As in the special relativistic formalism of the previous chapter, ρ is the local proper baryon rest-mass density which is simply related to the baryon number density n by $\rho = m_N n$. ϵ is used to denote the internal energy per gram of the fluid. It is convenient to follow two different spatial velocity fields. One is U_i, the spatial components of the covariant four-velocity. The other is V^i, the contravariant coordinate matter three-velocity. V^i is related to the four-velocity

$$V^i = \frac{U^i}{U^t} = \frac{\gamma^{ij} U_j}{U^t} - \beta^i, \tag{3.4}$$

where a shift vector β^i is often chosen to minimize coordinate fluid motion with respect to the shifting ADM grid and to keep the fluid coordinates in step with the metric coefficients.

Similar to the special relativity discussion of Chapter 2, we can introduce a Lorentz-like variable

$$W \equiv \alpha U^t, \tag{3.5}$$

where now the gravitational time dilation is accounted for in the extra factor α. The spatial three-velocity becomes,

$$V^i = \frac{U^i}{U^t} = \frac{\alpha \gamma^{ij} U_j}{W} - \beta^i. \tag{3.6}$$

As in the previous chapters, it is again convenient to remove the pressure as a variable by introducing an equation of state index

$$\Gamma = 1 + \frac{P}{\rho \epsilon}. \tag{3.7}$$

Next, we introduce the same Lorentz-contracted state variables [5] as in the special relativity discussion in Chapter 2. Using these it becomes possible to write the relativistic hydrodynamic equations in a form which is similar to the Newtonian counterparts.

The coordinate baryon mass density is written,

$$D = W\rho, \tag{3.8}$$

the coordinate internal energy density,

$$E = W\rho\epsilon, \tag{3.9}$$

and the covariant momentum density,

$$S_\mu = (D + \Gamma E)U_\mu. \tag{3.10}$$

We begin with the hydrodynamic equations for a perfect fluid. (We consider an imperfect fluid in the next section.) For a perfect fluid the energy momentum tensor only differs from the special relativistic version by the replacement of $g_{\mu\nu}$ for $\eta_{\mu\nu}$. Thus, we write,

$$T_{\mu\sigma} = (\rho + \rho\epsilon + P)U_\mu U_\sigma + Pg_{\mu\sigma}. \tag{3.11}$$

For general relativistic hydrodynamics it is convenient to work with the mixed form,

$$T^\nu{}_\mu = g^{\sigma\nu}T_{\mu\sigma} = (\rho + \rho\epsilon + P)U_\mu U^\nu + P\delta^\nu{}_\mu, \tag{3.12}$$

which in terms of our new state variables can also be written

$$T^\nu{}_\mu = \frac{S_\mu S^\nu}{S^t} + P\delta^\nu{}_\mu. \tag{3.13}$$

3.2.2 Equations of motion

With our definitions in place, it is now possible to impose the constraint conditions on the stress energy tensor and state variables. The simplest condition is that of continuity, which in general relativity becomes

$$\left(\rho U^\mu\right)_{;\mu} = \left(DV^\mu\right)_{;\mu} = \frac{1}{\sqrt{g}}\frac{\partial}{\partial x^\mu}\left(\sqrt{g}DV^\mu\right) = 0. \qquad (3.14)$$

Baryon number conservation then takes the form

$$\dot{D} + D\frac{\dot{\gamma}}{\gamma} + \frac{1}{\gamma}\frac{\partial}{\partial x^i}(\gamma DV^i) = 0, \qquad (3.15)$$

where $\dot{D}$ denotes differentiation with respect to coordinate time. Here, we see that a difference between general and special relativity enters through the appearance of terms containing $\gamma \equiv \sqrt{\det(\gamma_{ij})}$. These γ terms can be thought of as factors necessary to maintain proper coordinate volume. In addition, there will be gravitational acceleration terms.

The vanishing of the spatial components of the divergence of the energy momentum tensor,

$$T^\mu{}_{i;\mu} = 0, \qquad (3.16)$$

leads to the general relativistic momentum equation,

$$\frac{1}{\gamma}\frac{\partial(S_i\gamma)}{\partial t} + \frac{1}{\gamma}\frac{\partial(S_iV^j\gamma)}{\partial x^j} + \alpha\frac{\partial P}{\partial x^i} + \frac{1}{2}\frac{\partial g^{\alpha\beta}}{\partial x^i}\frac{S_\alpha S_\beta}{S^t} = 0. \qquad (3.17)$$

The first two terms are advection terms familiar from Newtonian fluid mechanics. The latter two terms comprise the pressure and gravitational forces. We can then expand the gravitational acceleration into individual contributions from metric variables. Thus, the momentum equation becomes

$$\dot{S}_i + S_i\frac{\dot{\gamma}}{\gamma} + \frac{1}{\gamma}\frac{\partial}{\partial x^j}(S_iV^j\gamma) + \alpha\frac{\partial P}{\partial x^i} - S_j\frac{\partial \beta^j}{\partial x^i}$$

$$+ (D+\Gamma E)\left(W\frac{\partial \alpha}{\partial x^i} + \frac{U_kU_j}{2W}\frac{\partial \gamma^{jk}}{\partial x^i}\right) = 0. \qquad (3.18)$$

Finally, the energy equation that we will use is derived by projecting the divergence of $T^{\mu\nu}$,

$$\left(U_\mu T^{\mu\nu}\right)_{;\nu} = 0. \qquad (3.19)$$

In our notation this becomes

$$\dot{E} + E\frac{\dot{\gamma}}{\gamma} + \frac{1}{\gamma}\frac{\partial}{\partial x^j}(EV^j\gamma) + \frac{P}{\gamma}\left[\frac{\partial(W\gamma)}{\partial t} + \frac{\partial}{\partial x^j}(WV^j\gamma)\right] = 0. \qquad (3.20)$$

Equivalently, we can write $P = (\Gamma - 1)E/W$ and collect the terms with $(\dot{\gamma}/\gamma)$ to yield an equation which is easier to solve numerically,

$$\dot{E} + \Gamma E \frac{\dot{\gamma}}{\gamma} + \frac{1}{\gamma}\frac{\partial}{\partial x^i}(EV^i\gamma) + (\Gamma - 1)E\left[\frac{\dot{W}}{W} + \frac{1}{\gamma W}\frac{\partial}{\partial x^i}(WV^j\gamma)\right] = 0.$$

$$(3.21)$$

3.2.3 Viscosity and heat flow

As in special relativity we can extend these equations to the form of the Navier–Stokes equations by adding an additional stress term $\Sigma_{\mu\nu}$ to $T_{\mu\nu}$.

$$T^{\mu\nu} = (\rho(1 + \epsilon) + P)U^\mu U^\nu + g^{\mu\nu}P + \Sigma^{\mu\nu}, \qquad (3.22)$$

where the viscous stress tensor $\Sigma^{\mu\nu}$ is as defined in Eq. (2.27). Now $\Sigma_{\mu\nu}$ can be boosted by a Lorentz transformation into the Eulerian frame. Additional terms must then be added to the momentum and energy equations. On the right side of the momentum equation (3.18) we add

$$= -\frac{1}{\sqrt{g}}\frac{\partial}{\partial x^\nu}\left(\sqrt{g}\Sigma_i^\nu\right) + \Gamma_{i\nu}^\lambda\Sigma_\lambda^\nu, \qquad (3.23)$$

and to the right-hand side of the energy equation ((3.20) or (3.21)) we add

$$= \left(U^\mu\Sigma^\nu{}_\mu\right)_{;\nu} - \Sigma^\nu{}_\mu U^\mu{}_{;\nu}. \qquad (3.24)$$

3.2.4 Grid velocities

To make the grid follow the fluid motion better, we introduce grid velocities, V_g^i. (Note that this is not equivalent to introducing a shift vector.) The perfect fluid equations are then:

$$\dot{D} + D\frac{\dot{\gamma}}{\gamma} + \frac{1}{\gamma}\frac{\partial}{\partial x^i}\left(\gamma D(V^i - V_g^i)\right) + \frac{D}{\gamma}\frac{\partial}{\partial x^i}\left(\gamma V_{g.}^i\right) = 0, \qquad (3.25)$$

$$\dot{S}_i + S_i\frac{\dot{\gamma}}{\gamma} - \frac{1}{\gamma}\frac{\partial}{\partial x^j}\left(S_i(V^j - V_g^i)\gamma\right) + \frac{S_i}{\gamma}\frac{\partial}{\partial x^i}\left(\gamma V_g^i\right) + \alpha\frac{\partial P}{\partial x^i}$$

$$- S_j\frac{\partial\beta^j}{\partial x^i} + (D + \Gamma E)\left(W\frac{\partial\alpha}{\partial x^i} + \frac{U_k U_j}{2W}\frac{\partial\gamma^{jk}}{\partial x^i}\right) = 0, \qquad (3.26)$$

$$\dot{E} + \Gamma E \frac{\dot{\gamma}}{\gamma} + \frac{1}{\gamma}\frac{\partial}{\partial x^i}\left(E(V^i - V_g^i)\gamma\right) + \frac{\Gamma E}{\gamma}\frac{\partial}{\partial x^i}\left(\gamma V_g^i\right)$$

$$+ (\Gamma - 1)E\left[\frac{\dot{W}}{W} + \frac{1}{\gamma W}\frac{\partial}{\partial x^i}\left(W(V^i - V_g^i)\gamma\right)\right] = 0. \qquad (3.27)$$

Note that V_g^i has spatial components only.

3.3 Difference equations

3.3.1 General relativistic hydrodynamics in one dimension

We will first discuss the difference between the special relativistic difference equations as solved in Chapter 2 and their general relativistic one-dimensional counterparts. Afterward, we will outline how one takes the one-dimensional algorithms into three dimensions.

3.3.2 Operator splitting

Here, as in Chapter 2, we make use of operator splitting. That is, quantities are advanced in time by evaluating terms in the evolution equations one after another. As in Chapter 2, the order in which these equations are used is important. For a general relativistic calculation we have found that the following order works well:

1. acceleration by pressure gradients
2. viscosity
3. velocities
4. pressure, heating, cooling
5. advection
6. velocities again
7. time step
8. grid
9. metric calculation
10. $\dot{\gamma}/\gamma$ proper volume terms
11. output and post processing when appropriate.

3.3.3 Proper volume terms

First we consider terms involving $(\dot{\gamma}/\gamma)$ in Eqs. (3.25), (3.26), and (3.27). When the field variables are updated, a new volume factor γ is calculated. The state variables are then updated as

$$D' = D\frac{\gamma}{\gamma'}, \qquad E' = E\left(\frac{\gamma}{\gamma'}\right)^{\Gamma}, \qquad S' = S\frac{\gamma}{\gamma'}, \qquad (3.28)$$

where the prime denotes a new value and the coordinate index on the momenta is dropped as we are considering a system with only one dynamic dimension. Note that this conserves these quantities exactly as the metric is updated.

3.3.4 Advection in one dimension

Next we treat the advection terms, e.g.

$$\frac{1}{\gamma}\frac{\partial}{\partial x}\left(\gamma DV\right). \tag{3.29}$$

To simplify the difference equation, we first make the replacement,

$$\hat{D} = \gamma D, \qquad \hat{E} = \gamma E, \qquad \hat{S} = \gamma S. \tag{3.30}$$

Then, the advection parts of Eqs. (3.25), (3.26), and (3.27) become,

$$\frac{\partial \hat{D}}{\partial t} + \frac{\partial}{\partial x}\left(\hat{D}(V - V_g)\right) = 0, \tag{3.31}$$

$$\frac{\partial \hat{E}}{\partial t} + \frac{\partial}{\partial x}\left(\hat{E}(V - V_g)\right) = 0, \tag{3.32}$$

$$\frac{\partial \hat{S}}{\partial t} + \frac{\partial}{\partial x}\left(\hat{S}(V - V_g)\right) = 0. \tag{3.33}$$

These advection terms can then be treated numerically (using the velocities $(V - V_g)$) exactly as they were treated in Chapter 2. The new D, E, and S, are then restored to their proper values by dividing by γ.

3.3.5 Grid velocity advection

Next, the grid velocity advection is evaluated by forming the divergence of (γV_g). Henceforth, where simple spatial derivatives are used we will just use $\partial/\partial x$ instead of its finite differenced form.

Let,

$$\nabla V_g \equiv \frac{1}{\gamma}\frac{\partial}{\partial x}\left(\gamma V_g\right). \tag{3.34}$$

Then we have,

$$D' = De^{-\nabla V_g \Delta t}, \tag{3.35}$$

$$E' = Ee^{-\Gamma \nabla V_g \Delta t}, \tag{3.36}$$

$$S' = Se^{-\nabla V_g \Delta t}. \tag{3.37}$$

The last term in the energy equation (3.27) is of the form

$$\dot{E} = -(\Gamma - 1)E\left[\frac{\dot{W}}{W} + \frac{1}{\gamma W}\frac{\partial}{\partial x^i}\Big(W(V - V_g)\gamma\Big)\right].\qquad(3.38)$$

The $(\dot{W}/W)$ term is evaluated exactly as in Chapter 2 (cf. Eq. (2.88)), i.e.

$$E' = E\left(\frac{W}{W'}\right)^{\Gamma-1}.\qquad(3.39)$$

The remaining term becomes

$$E' = E\exp\left\{-\frac{(\Gamma - 1)\Delta t}{\gamma W}\frac{\partial}{\partial x}\Big(W(V - V_g)\gamma\Big)\right\}.\qquad(3.40)$$

The W inside the derivative is evaluated at the zone nodes and the W outside the derivative is an average of the nodal values. The γ outside is zonal centered so that the γ inside the derivative is taken as an average of the zonal γ to make nodal γ.

3.3.6 Pressure acceleration

Next, the pressure acceleration is straightforward. The only difference with special relativity (Section 2.4.12) is the presence of the α factor in front of the gradient. We update the momentum density in a straightforward way,

$$S' = S - \Delta t\,\bar{\alpha}\,\frac{\partial P}{\partial x},\qquad(3.41)$$

where in one dimension, $\bar{\alpha} = (\alpha(i) + \alpha(i-1))/2$ is a node-centered average value for α.

3.3.7 Metric acceleration

Finally, the metric acceleration terms are simply,

$$\dot{S} - S\frac{\partial\beta}{\partial x} + (D + \Gamma E)W\left(\frac{\partial\alpha}{\partial x} + \frac{U^2}{W^2}\frac{\partial\gamma^{xx}}{\partial x}\right) = 0.\qquad(3.42)$$

These can be applied in two steps:

$$S' = Se^{\Delta t\partial\beta/\partial x},\qquad(3.43)$$

and

$$S'' = S' - \Delta t\,(D + \Gamma E)W\left(\frac{\partial\alpha}{\partial x} + \frac{U^2}{W^2}\frac{\partial\gamma^{xx}}{\partial x}\right).\qquad(3.44)$$

The variable centering for this is again straightforward. It simply requires that $(D + \Gamma E)$ be averaged over adjacent zones.

3.4 Multi-dimensional difference equations

The finite difference equations begin to appear more complicated as one goes to higher dimensions. The main difference with the one-dimensional form described above is that the centering of different scalar quantities and vector components must be done carefully to maintain accuracy during the simulation. Hence, we begin with a discussion of optimum centering for the variables, and then discuss the numerical solution of the hydrodynamics.

3.4.1 The grid

The locations of the coordinate variables for a three-dimensional calculation are shown in Figures 3.1 and 3.2.

We introduce a finite grid of points $x(i), y(j), z(k)$, labeled by integers i, j, k. x, y, z are not necessarily Cartesian at this point. However, we will take the coordinates to be locally orthogonal. We can thus think of a cube to represent the space for one zone as depicted in Figure 3.1. $x_a(i, j, k)$, $y_a(i, j, k)$, and $z_a(i, j, k)$ are the coordinates of the lower left corner of the grid. Let $x_b(i, j, k)$ be between $x_a(i, j, k)$ and $x_a(i+1, j, k)$ weighted such that x_a and x_b form smooth arrays in space. Similarly we define $y_b(i, j, k)$ and $z_b(i, j, k)$.

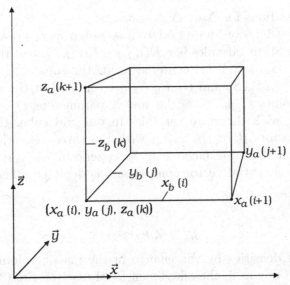

Fig. 3.1. Schematic representation of the cube representing one grid zone.

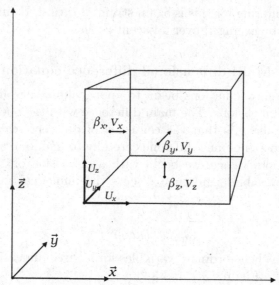

Fig. 3.2. Schematic representation of the centering of vectors in the spatial grid.

Next we define differences in coordinate positions as

$$\Delta x_a(i,j,k) = x_b(i,j,k) - x_b(i-1,j,k), \qquad (3.45)$$

$$\Delta x_b(i,j,k) = x_a(i+1,j,k) - \Delta x_a(i,j,k), \qquad (3.46)$$

with similar relations for $\Delta y_a,\ \Delta y_b,\ \Delta z_a,\ \Delta z_b$.

Having selected a coordinate grid we now assign spatial positions to the fluid variables. State variables like $D(i,j,k)$, $E(i,j,k)$, and $W(i,j,k)$ will be centered at (x_b, y_b, z_b), i.e. centered inside the zone cube. Differential quantities like S_x, S_y, S_z and U_x, U_y, U_z are centered at the nodes of the cube, i.e. at points x_a, y_a, z_a. Scalar metric parameters, i.e. α, γ, γ_{ij} are centered along with the state variables in the grid cube. However, the metric shift vectors $\beta_x, \beta_y, \beta_z$ along with the three velocities V_x, V_y, V_z are centered on the cube faces, e.g. β_x appears at x_a, y_b, z_b. The centering of various spatial vector components is depicted schematically in Figure 3.2.

3.4.2 Advection

The change of densities by the change in the metric volume factors is made the same way as in the one-dimensional case (cf. Eq. (3.28)) at the time the new spatial metric is evaluated.

The advection terms for D and E are performed by a coordinate transformation to $\tilde{D}$ and $\tilde{E}$ in the same manner as for the one-dimensional case. For multi-dimensional problems, the advection is made in each direction separately. There is, however, a possible centering error arising from the order of advecting in the different directions sequentially. To overcome this, the order of advection is alternated, i.e. first advection is done in the x^1, x^2, x^3 directions for half a time step. Then the calculation is done in the order x^3, x^2, x^1 in the second half step. If only two directions are being considered, a similar mixing of ordering is used in those two directions.

On each zonal face, the fluxes of inertial density $(D+\Gamma E)$ are evaluated in the same manner as for special relativity (see Section 2.4.16). Call the flux of inertial density $(D + \Gamma E)$ in the direction i as f_i. Now think of a cube surrounding the node where S_i (and U_i on Figure 3.2) resides. Average the f_i on the node-centered faces of that cube. This gives the flux of inertial mass in and out of the cube. Next, it is necessary to find the four-velocity components $\tilde{U}_j$ to be associated with the mass fluxes. The line joining nodes is in the center of these momentum faces so that the $\tilde{U}_j$ are calculated on each line by the same method as used in the special relativity discussions (Section 2.4.16, Eqs. (2.127), (2.128)). The fluxes across the faces are then formed by the products $f_i\tilde{U}_j$. Hence,

$$S_i' = S_i + \sum_j f_j \tilde{U}_i. \tag{3.47}$$

The j summation is over all six faces of the cube.

3.4.3 Advection in curvilinear coordinates

In the case of curved coordinates it is best to advect Cartesian specific momenta rather than the coordinate momenta. For example, in cylindrical coordinates with momentum components, S_r, S_z, S_θ, the corresponding Cartesian values are:

$$S_x = S_r \cos\theta - \frac{S_\theta \sin\theta}{r}, \tag{3.48}$$

$$S_y = S_r \sin\theta + \frac{S_\theta \cos\theta}{r}, \tag{3.49}$$

$$S_z = S_z, \tag{3.50}$$

and

$$U_x = U_r \cos\theta - \frac{U_\theta \sin\theta}{r}, \tag{3.51}$$

$$U_y = U_r \sin\theta + \frac{U_\theta \cos\theta}{r}, \tag{3.52}$$

$$U_z = U_z. \tag{3.53}$$

Once these are evaluated, the zone-centered four-velocities $\tilde{U}_x, \tilde{U}_y, \tilde{U}_z$ for the momentum transfer are then calculated as in Eqs. (2.128) and (2.129). Then, after calculating the new values for S_x and S_y, the momentum components are transformed back to the r, z, θ frame.

Note that this process has already taken into account the centrifugal forces that would arise from the term $(\partial g^{\theta\theta}/\partial r)$. This is compensated for by replacing $(\partial g^{\theta\theta}/\partial r)$ with

$$\frac{1}{r}\frac{\partial}{\partial r}\left(rg^{\theta\theta}\right)$$

in the gravitational field acceleration terms.

After all advection steps have been completed the $\hat{D}, \hat{E}$, and $\hat{S}_i$ are divided by the appropriate γ factor to restore D, E, S_i as proper coordinate densities.

3.4.4 Pressure terms

Next we evaluate the multi-dimensional PdV terms,

$$\dot{E} = -(\Gamma - 1)E\left[\frac{\dot{W}}{W} + \frac{1}{\gamma W}\frac{\partial}{\partial x^i}\left((W\gamma(V^i - V_g^i))\right)\right]. \tag{3.54}$$

The terms involving $\dot{W}$ and $\partial W/\partial x^i$ are treated later in the velocity update, as W is derived from the four-velocity. Thus, here we only update the term

$$\dot{E} = -(\Gamma - 1)E\left[\frac{\partial}{\partial x^i}\left((\gamma(V^i - V_g^i))\right)\right]. \tag{3.55}$$

We begin by averaging γ to form face-centered values of the curvature-scaled velocity, $V_g^i\gamma$ (face-centered). Then, the derivatives dV_g^i/dx^i can be taken along the three directions. The energy is then updated, i.e.

$$E' = E\exp\left\{-\frac{(\Gamma - 1)}{\gamma}\vec{\nabla}\cdot(\vec{V}_\mathbf{g}\gamma)\Delta t\right\}. \tag{3.56}$$

Here, a sketch of how this is done in FORTRAN is helpful. We will introduce a simple compact notation whereby *do i,j,k* denotes a loop over all the grid indices. The reader must determine from the context whether the indices begin and end with the first and last indices, or one or two indices from either end. The end of all three loops will be denoted with a single *enddo*, though this would be repeated for each index in a real code.

```
do i, j, k
```
$$A_x(i,j,k) = (V_x(i,j,k) - V_x^g(i,j,k)) * (\gamma(i-1,j,k) + \gamma(i,j,k))/2$$
$$A_y(i,j,k) = (V_y(i,j,k) - V_y^g(i,j,k)) * (\gamma(i,j-1,k) + \gamma(i,j,k))/2$$
$$A_z(i,j,k) = (V_z(i,j,k) - V_z^g(i,j,k)) * (\gamma(i,j,k-1) + \gamma(i,j,k))/2$$
```
enddo
```

```
do i, j, k
```
$$q_x = (A_x(i+1,j,k) - A_x(i,j,k))/\Delta x_b(i)$$
$$q_y = (A_y(i,j+1,k) - A_y(i,j,k))/\Delta y_b(j)$$
$$q_z = (A_z(i,j,k+1) - A_z(i,j,k))/\Delta z_b(k)$$
$$divv = (q_x + q_y + q_z)$$
$$E(i,j,k) = E(i,j,k) * exp[-(\Gamma(i,j,k)-1) * divv * \Delta t/\gamma(i,j,k)]$$
```
enddo
```

In the above, we have assumed as usual that the adiabatic index Γ is a slowly varying function of D and E and t.

3.4.5 Velocity update

Next we give the method of calculating the velocities. As in Chapter 2, at the same time as velocities are recalculated, the changes of E due to W changes are concurrently calculated. As in special relativity (Eqs. (2.83) and (2.84)) we calculate the covariant four-velocities U_i from the momentum density S_i. To do this we must first evaluate an inertial density

$$\sigma(i,j,k) = D(i,j,k) + \Gamma(i,j,k)E(i,j,k), \tag{3.57}$$

then divide the momentum density S_i by the appropriately node-centered inertial density $\bar{\sigma}$,

$$U_x(i,j,k) = S_x(i,j,k)/\bar{\sigma}(i,j,k). \tag{3.58}$$

To show how this is done in three dimensions, we here sketch some of the relevant FORTRAN.

```
do i, j, k
```
$$x1 = \left(\sigma(i,j,k) * \Delta x_b(i) + \sigma(i-1,j,k) * dx_b(i-1)\right) * \Delta y_b(j) * \Delta z_b(k)$$

$$x2 = \left(\sigma(i,j-1,k) * \Delta x_b(i) + \sigma(i-1,j-1,k) * \Delta x_b(i-1)\right)$$
$$* \Delta y_b(j-1) * \Delta z_b(k)$$

$$x3 = \Big(\sigma(i,j,k-1)*\Delta x_b(i) + \sigma(i-1,j,k-1)*dx_b(i-1)\Big)$$
$$\quad *\Delta y_b(j)*\Delta z_b(k-1)$$
$$x4 = \Big(\sigma(i,j-1,k-1)*\Delta x_b(i) + \sigma(i-1,j-1,k-1)*\Delta x_b(i-1)\Big)$$
$$\quad *\Delta y_b(j-1)*\Delta z_b(k-1)$$
$$qq = 8.0*\Delta x_a(i)*\Delta y_a(j)*\Delta z_a(k)/(x1+x2+x3+x4)$$
$$U_x(i,j,k) = S_x(i,j,k)*qq$$
$$U_y(i,j,k) = S_y(i,j,k)*qq$$
$$U_z(i,j,k) = S_z(i,j,k)*qq$$
enddo

Next, the face-centered contravariant velocities V^x, V^y, V^z and the corresponding face-centered W factors Wx, Wy, Wz are calculated. Note that the notation Wx refers to centering along the x face (cf. Figure 3.2 x–z plane) of the nodal grid cube. It is not a vector component. With appropriately face-centered W factors, we use Eq. (3.4) to find V^i,

$$V^i = \frac{U^i}{U^t} = \frac{\alpha\gamma^{ij}\bar{U}_j}{W} - \beta^i, \tag{3.59}$$

where the bar denotes a face-centered average of U_i. We next illustrate how to generate face-centered x_a quantities. Averaging on the other faces can be achieved by an obvious extension.

The face-centered four-velocities are found by averaging over the four nodes around each face:

do i, j, k

$$\bar{U}_x(x_a) = \Big(U_x(i,j,k) + U_x(i,j+1,k) + U_x(i,j,k+1) + U_x(i,j+1,k+1)\Big)/4$$

$$\bar{U}_y(x_a) = \Big(U_y(i,j,k) + U_y(i,j+1,k) + U_y(i,j,k+1) + U_y(i,j+1,k+1)\Big)/4$$

$$\bar{U}_z(x_a) = \Big(U_z(i,j,k) + U_z(i,j+1,k) + U_z(i,j,k+1) + U_z(i,j+1,k+1)\Big)/4$$

where here the (x_a) reminds us that the average was made on the x_a face. Henceforth, however, we drop this notation. Next we can form the contravariant $\bar{U}^i$. For this example we will assume that γ^{ij} is diagonal as is the case for most applications discussed in this book. First the face-centered γ^{ii} is found,

$$\bar{\gamma}^{xx}(i,j,k) = (\gamma^{xx}(i,j,k) + \gamma^{xx}(i-1,j,k))/2$$
$$\bar{\gamma}^{yy}(i,j,k) = (\gamma^{yy}(i,j,k) + \gamma^{yy}(i-1,j,k))/2$$
$$\bar{\gamma}^{zz}(i,j,k) = (\gamma^{zz}(i,j,k) + \gamma^{zz}(i-1,j,k))/2$$

then the contravariant $\bar{U}^i$ is formed

$$\bar{U}^x = \bar{U}_x * \bar{\gamma}^{xx}$$
$$\bar{U}^y = \bar{U}_y * \bar{\gamma}^{yy}$$
$$\bar{U}^z = \bar{U}_z * \bar{\gamma}^{zz}$$

The face-centered lapse function is simply

$$\bar{\alpha} = (\alpha(i,j,k) + \alpha(i-1,j,k))/2$$

The x_a-centered Lorentz-like factor W_{x_a} can then be determined

$$W_{x_a}(i,j,k) = \sqrt{(1.0 + \bar{U}_x\bar{U}^x + \bar{U}_y\bar{U}^y + \bar{U}_z\bar{U}^z)}$$

and the face-centered x component of the three-velocity determined

$$V^x(i,j,k) = \bar{\alpha}(i,j,k) * [\bar{U}^x/W_{x_a}(i,j,k)] - \beta^x(i,j,k)$$
enddo

The above loop is then repeated for the other two directions.

3.4.6 Energy PdV velocity terms

To update the velocity dependent PdV terms in the energy equation, we must first evaluate a cube-centered W and four-velocities.

The cube-centered four-velocities $\tilde{U}_i$ are formed by averaging around the eight corners of the cube

do i,j,k

$$\tilde{U}_x = [U_x(i,j,k) + U_x(i+1,j,k) + U_x(i,j+1,k) + U_x(i,j,k+1)$$
$$+ U_x(i+1,j,k+1) + U_x(i,j+1,k+1) + U_x(i+1,j+1,k)$$
$$+ U_x(i+1,j+1,k+1)]/8$$
$$\tilde{U}_y = [U_y(i,j,k) + U_y(i+1,j,k) + U_y(i,j+1,k) + U_y(i,j,k+1)$$
$$+ U_y(i+1,j,k+1) + U_y(i,j+1,k+1) + U_y(i+1,j+1,k)$$
$$+ U_y(i+1,j+1,k+1)]/8$$
$$\tilde{U}_z = [U_z(i,j,k) + U_z(i+1,j,k) + U_z(i,j+1,k) + U_z(i,j,k+1)$$
$$+ U_z(i+1,j,k+1) + U_z(i,j+1,k+1) + U_z(i+1,j+1,k)$$
$$+ U_z(i+1,j+1,k+1)]/8$$

Then the contravariant form is computed

$$\tilde{U}^x = \tilde{U}_x * \gamma^{xx}(i,j,k)$$
$$\tilde{U}^y = \tilde{U}_y * \gamma^{yy}(i,j,k)$$
$$\tilde{U}^z = \tilde{U}_z * \gamma^{zz}(i,j,k)$$

and the new zone-centered W evaluated

$$W_{new}(i,j,k) = \sqrt{(1.0 + (\tilde{U}_x * \tilde{U}^x + \tilde{U}_y * \tilde{U}^y + \tilde{U}_z * \tilde{U}^z)}$$

With W properly centered, we can now update the $\dot{W}$ term in the energy equation using the face-centered W

$\cdot E(i,j,k) = E(i,j,k) * (W(i,j,k)/W_{new})^{(\Gamma(i,j,k)-1)}$
$W(i,j,k) = W_{new}$
enddo

Finally, we update the $P(V^i - V^i_g)\partial W/\partial x^i$ term in the energy equation as follows:

do i,j,k

$$qx = \left[V^x(i,j,k) + V^x(i+1,j,k)) - V^x_g(i,j,k) - V^x_g(i+1,j,k)\right]$$
$$* \left(W_{x_a}(i+1,j,k) - W_{x_a}(i,j,k)\right) * \Delta t/2\Delta x_b(i)$$

$$qy = \left[V^y(i,j,k) + V^y(i,j+1,k)) - V^y_g(i,j,k) - V^y_g(i,j+1,k)\right]$$
$$* \left(W_{y_a}(i,j+1,k) - W_{y_a}(i,j,k)\right) * \Delta t/2 * \Delta y_b(j)$$

$$qz = \left[V^z(i,j,k) + V^z(i,j,k+1)) - V^z_g(i,j,k) - V^z_g(i,j,k+1)\right]$$
$$* \left(W_{z_a}(i,j,k+1) - W_{z_a}(i,j,k)\right) * \Delta t/2\Delta z_b(k)$$

$\nabla V = (qx + qy + qz)/W(i,j,k)$
$E(i,j,k) = E(i,j,k) * e^{(-\Gamma(i,j,k)-1)*\nabla V}$
enddo

3.5 Grid calculation

The grid motion is chosen according to the problem at hand, but the two most common ways to treat the grid motion are to select either uniform zoning, i.e. equal spaced zones in each direction, or exponentially varying zonal space increments, i.e. $dx(i+1) = dx(i) * k$, where k is a constant. Uniform zones give the most accurate results with the difference equations in the form given. With exponentially varying zones, no appreciable error occurs as long as $0.95 < k < 1.05$. However, more extreme values of k must be used with caution and are generally not a good choice.

The grid is moved with these restrictions so as to follow the fluid flow as best as possible. In one-dimensional problems we can let $V_g \to V$ and effectively have a Lagrangian gauge.

Having selected the three components of the grid velocity V_g^x, V_g^y, V_g^z we then want to treat the terms,

$$\dot{E} = -\frac{E}{\gamma}\frac{\partial(\gamma V_g^i)}{\partial x^i}, \tag{3.60}$$

$$\dot{D} = -\frac{D}{\gamma}\frac{\partial(\gamma V_g^i)}{\partial x^i}, \tag{3.61}$$

$$\dot{S}_j = -\frac{S_j}{\gamma}\frac{\partial(\gamma V_g^i)}{\partial x^i}. \tag{3.62}$$

When the grid is moved we should also move the metric functions which would result, for example, in $\dot{\gamma} + V_g^i(\partial\gamma/\partial x^i) = 0$. Instead of changing γ here we should resolve Eqs. (3.60), (3.61), and (3.62) in a way which accounts for the changing coordinate volume on the grid. The implied γ term is taken care of when the metrics are recalculated.

To do this we form a new grid and take the ratio of the old coordinate volumes to the new coordinate volumes

$$Ratio = \frac{\left(dx_b(i)*dy_b(j)*dz_b(k)\right)_{old}}{\left(dx_b(i)*dy_b(j)*dz_b(k)\right)_{new}}. \tag{3.63}$$

Then,

$$D(i,j,k) = D(i,j,k)*Ratio. \tag{3.64}$$

The S_i are similarly updated, but using a ratio of coordinate volumes determined at the nodes, i.e. dx_a, dy_a, dz_a instead of dx_b, dy_b, dz_b in Eq. (3.63).

From the pressure part of $\dot{E}$ we also have a term left over by our splitting of terms

$$\frac{PW}{\gamma}\frac{\partial V_g^i}{\partial x^i},$$

which we include with the above formula for $\dot{E}$ by writing

$$\dot{E} = -\frac{\Gamma E}{\gamma}\frac{\partial(\gamma V_g^i)}{\partial x^i}. \tag{3.65}$$

Thus the updated energy after changing the grid is

$$E' = E*Ratio^\Gamma, \tag{3.66}$$

or

$$E(i,j,k) = E(i,j,k)*Ratio^\Gamma. \tag{3.67}$$

3.5.1 Pressure acceleration

In order to solve the pressure acceleration term in multiple dimensions, $\dot{S}_i = -\alpha \partial P/\partial x^i$ appropriate averages must again be made. In three dimensions we form an average of α over the eight zones surrounding the node where S_i resides. Then an average of $P = (\Gamma - 1)E/W$ is made over the four zones bordering the line between, for example, $S_x(i,j,k)$ and $S_x(i+1,j,k)$. Then S_x is advanced by

$$S_x(i,j,k) = S_x(i,j,k) - \bar{\alpha} * \left(\frac{\bar{P}(i,j,k) - \bar{P}(i-1,j,k)}{[x_b(i,j,k) - x_b(i-1,j,k)]} \right) * \Delta t. \quad (3.68)$$

3.5.2 Metric acceleration

The relevant terms from Eq. (3.18) are:

$$S_i = -S_j \frac{\partial \beta^j}{\partial x^i} + (D + \Gamma E)\left(W \frac{\partial \alpha}{\partial x^i} + \frac{U_k U_j}{W} \frac{\partial \gamma^{ij}}{\partial x^i} \right). \quad (3.69)$$

Here, the multi-dimensional algorithms are slightly different than in one dimension.

3.5.3 Shift vector acceleration

The shift vector components β^i are centered on faces (cf. Figure 3.2). We illustrate the difference equations for S_x only. First, line-centered averages for the β^i components are produced.

do i,j,k

$$\bar{\beta}^x = \Big(\beta^x(i,j,k) + \beta^x(i,j-1,k) + \beta^x(i,j,k-1) + \beta^x(i,j-1,k-1)$$
$$+ \beta^x(i+1,j,k) + \beta^x(i+1,j-1,k) + \beta^x(i+1,j,k-1)$$
$$+ \beta^x(i+1,j-1,k-1) \Big)/8$$

$$\bar{\beta}^y = \Big(\beta^y(i,j,k) + \beta^y(i,j,k-1) \Big)/2$$

$$\bar{\beta}^z = \Big(\beta^z(i,j,k) + \beta^y(i,j-1,k) \Big)/2$$

enddo

Then, the momentum densities are updated.

do i,j,k

$$S_x(i,j,k) = S_x(i,j,k)$$
$$+ S_x(i,j,k) * \Delta t * \left(\Big(\bar{\beta}^x(i,j,k) - \bar{\beta}^x(i-1,j,k) \Big)/\Delta x_a(i) \right)$$

$$+ S_y(i,j,k) * \Delta t * \left(\left(\bar{\beta}^y(i,j,k) - \bar{\beta}^y(i-1,j,k) \right) / \Delta y_a(j) \right.$$

$$+ S_z(i,j,k) * \Delta t * \left(\left(\bar{\beta}^z(i,j,k) - \bar{\beta}^z(i-1,j,k) \right) / \Delta z_a(k) \right.$$

enddo

3.5.4 Lapse and three-metric acceleration

The accelerations by the metric terms α and γ^{ij} are similar. Since the metric functions are centered at the same place as the pressure, the same averaging is performed for these gravity terms as for the pressure acceleration.

Remember that if we put in a coordinate curvature term in the advection section, then it must be removed in the metric driven terms (cf. Section 3.4.3). This completes the hydrodynamic difference equations for three spatial dimensions.

3.6 Artificial viscosity

For most applications in astrophysics the real physical viscosity produces shock waves whose thickness is much smaller than the zone size used in numerical simulations. Therefore, an artificial viscosity is introduced (as described in Chapter 2) which operates on the scale of the zone size. The artificial viscosity will operate on the derivatives of the velocities. It uses the method of calculating derivatives of the same nature as the monotonic derivatives introduced in the advection section. It is important to have artificial viscosity *not* acting where there are *not* real shock waves. The monotonic algorithm avoids most cases where spurious viscosity could arise.

First, form a zone-centered divergence of U^i. In obvious notation we write:

do i,j,k
$U_x = U_x(i,j,k) + U_x(i,j+1,k) + U_x(i,j,k+1) + U_x(i,j+1,k+1)$
$U_{x+1} = U_x(i+1,j,k) + U_x(i+1,j+1,k) + U_x(i+1,j,k+1)$
$\qquad + U_x(i+1,j+1,k+1)$
$U_y = U_x(i,j,k) + U_y(i+1,j,k) + U_y(i,j,k+1) + U_y(i+1,j,k+1)$
$U_{y+1} = U_y(i,j+1,k+1) + U_y(i+1,j+1,k) + U_y(i,j+1,k+1)$
$\qquad + U_y(i+1,j+1),k+1)$
$U_z = U_z(i,j,k) + U_z(i+1,j,k) + U_z(i,j+1,k) + v^z(i+1,j+1,k)$
$U_{z+1} = U_x(i,j,k+1) + U_x(i+1,j,k+1) + U_z(i,j+1,k+1)$
$\qquad + U_z(i+1,j+1),k+1)$

$$\nabla U(i,j,k) = \Big[(U_{x+1} - U_x)/\Delta x_b(i) + (U_{y+1} - U_y)/\Delta y_b(j)$$
$$+ (U_{y+1} - U_y)/\Delta z_b(k)\Big]/4$$

enddo

The quantity $\nabla U(i,j,k)$ will be used to test whether the artificial stress should exist in a zone. If $\nabla U(i,j,k) > 0$ then the artificial stress in that zone is set to zero.

Next, a node-centered W factor is calculated, $\bar{W} = \sqrt{1 + \gamma^{ij} U_j U_i}$. The γ^{ij} (which are zone centered) must be averaged to the node. We summarize here the calculation of the artificial stress arising in the x direction. Adopting the notation $U(i,j,k) \equiv U(i)$, for all j, k, then

do $i = 2, imm$
$U_{min} = min(U(i-1), U(i), U(i+1))$
$U_{max} = max(U(i-1), U(i), U(i+1))$
$Udix = min(U_{max} - U(i), U(i) - U_{min})$
if $((U(i+1) - U(i)) * (U(i) - U(i-1)) > 0)$ then
$Udiv = Udix$
else
$Udiv = 0$
endif
$$c(i) = \Big[dx_b(i) * (U(i) - U(i-1)) + dx_b(i-1) * (U(i+1) - U(i)) \Big]$$
$$\Big/ \Big[dxa(i) * (dx_b(i-1) + dx_b(i)) \Big]$$
$Ratio = 1.0$
if$(|c(i)| > 0)$ $Ratio = Udiv/c(i))$
$Ratio = min(Ratio, 1.0)$
enddo

do $i = 2, imm$
$$g(i) = \Big[\bar{W}(i,j,k) * dxa(i+1) + \bar{W}(i,j,k) * dxa(i) \Big]/dx_b(i)$$
$$f(i) = D(i,j,k) + D(i,j,k-1) + D(i,j-1.k) + D(i,j-1,k-1)$$
$$+ \Gamma(i,j,k) * E(i,j,k) + \Gamma(i,j,k-1) * E(i,j,k-1)$$
$$+ \Gamma(i,j-1,k) * E(i,j-1,k) + \Gamma(i,j-1,k-1) * E(i,j-1,k-1)$$
$$\delta = \Big[U(i+1) - U(i) - \Big(dx_b(i)/2 \Big) * \Big(c(i) + c(i+1)/g(i) \Big)$$

if $(\delta * (U(i+1) - U(i)) > 0)$ then
$\delta_m = \delta$
else

$$\delta_m = 0$$
endif

$$\delta_v = min(\delta_m, 0.0)$$
$$q_x(i, j, k) = f(i) * (Q * \delta_v - L * c_s)$$
$$t_x(i, j, k) = f(i) * \delta_v * \left[V^x(i, j, k) + V^x(i+1, j, k) - V^x_g(i) - V^{i+1}_g \right]$$

enddo

Here c_s is the speed of sound averaged over the line joining $V^x(i, j, k)$ and $V^x(i+1, j, k)$. The quantities Q and L are constants,

$$Q = dt * \left(\bar{\Gamma} + 1/2 \right)$$
$$L = dt = 1.5$$

The quantity $\bar{\Gamma}$ is an appropriate average for Γ for the problem at hand. The quantity $q_x(i, j, k)$ is the artificial stress due to motion in the x direction. q_y and q_z are formed in a similar manner in the other two directions. The quantity $t_x(i, j, k)$ we discuss below.

We then calculate a zone-centered average artificial scalar stress q by the following loop.

```
do i, j, k
```
$$a_x = q_x(i, j, k) + q_x(i, j, k+1) + q_x(i, j+1, k) + qz(i, j+1, k+1)$$
$$a_y = q_y(i, j, k) + q_y(i, j, k+1) + q_y(i+1, j, k) + qz(i+1, j, k+1)$$
$$a_z = q_z(i, j, k) + q_z(i, j+1, I) + q_z(i+1, j, k) + qz(i+1, j+1, k)$$
$$q(i, j, k) = (a_x + a_y + a_z)/4$$
if $((\partial U_l / \partial x^l) > 0.0)$ $q(i, j, k) = 0.0$
```
enddo
```

Now the acceleration is accomplished in a similar manner to the pressure acceleration. To conserve total momentum, the acceleration is taken to be (cf. Eq. (2.79))

$$\dot{S}_i = -\frac{1}{\gamma} \frac{\partial q}{\partial x^i}, \tag{3.70}$$

where γ and q are averaged as in the pressure force calculation described above.

The heating by the artificial stress is augmented by a term which arises from the dissipation of kinetic energy by advection which occurs in the steep velocity gradients present in a shock.

We now combine t_x, t_y, t_z as follows:

$$a = t_x(i, j, k) + t_x(i, j+1, k) + t_x(i, j, k+1) + t_x(i, j+1, k+1)$$
$$b = t_y(i, j, k) + t_y(i+1, j, k) + t_y(i, j, k+1) + t_y(i+1, j, k+1)$$
$$c = t_z(i, j, k) + t_z(i+1, j, k) + t_z(i, j, k+1) + t_z(i+1, j+1, k)$$

$$qt = q(i,j,k) - (a+b+c) * A/32$$
$$\text{if } ((\partial U_l / \partial x^l) < 0) \quad qt = 0$$

where A is a viscosity constant chosen to be 0.25. The heating of matter (cf. Eq. (2.81)) is then given by,

$$E(i,j,k) = E(i,j,k) + (q(i,j,k) - qt) * (\partial U_l / \partial x^l). \tag{3.71}$$

In the above we have constructed an artificial scalar stress q which assumes no rotation and no directional dependence. However, if the zone increments Δx, Δy, Δz are not comparable we must compute a one-dimensional stress in each direction. In this case, Eq. (3.70) involves three different q_i along the different directions.

3.7 Real viscosity and heat flow

Real viscosity has been put into numerical simulations both in special relativity (cf. Chapter 2, heavy ion collisions discussion) and in stellar collapse calculations (cf. Chapter 5). The numerical procedure is to form the velocity derivatives $(\partial U_i / x^j)$ and apply a Lorentz boost Λ_i^ν to the fluid frame so as to form

$$\frac{\partial \hat{U}_i}{\partial x^j} = \Lambda_i^\nu \Lambda_j^\mu \frac{\partial U_\nu}{\partial x^\mu}. \tag{3.72}$$

These are the velocity derivatives in the fluid frame. The deviatoric stress tensor is then evaluated in the fluid frame. This tensor is then boosted back to the Eulerian frame and the momentum and energy densities are advanced.

3.8 Time step

All of the algorithms presented above are performed in an explicit manner, i.e. sequentially. The size of the time step, dt, must be limited for both stability and accuracy. We impose the following method to calculate the time step.

The Courant stability criterion is that the zonal sound crossing times should be greater than the numerically applied time step. This criterion is carried out as follows. The sound speed in our notation is

$$c_s^2 = \sqrt{\Gamma E(\Gamma - 1)/(D + \Gamma E)}. \tag{3.73}$$

We then form for each zone:

$$A = min(dx_b(i), dy_b(j), dz_b(k))$$

$$B = c_s^2 \alpha^2 / A^2$$

$$C = \sqrt{max(B)}$$

Next, a restriction on the time step is required because the artificial viscosity is done explicitly. Viscosity is a diffusion-like process so we use an analogy to the stability requirements derived for the linear diffusion equation. Namely,

$$\Delta t < \frac{\Delta x^2}{4D},$$

where D is the diffusion coefficient. Let,

$$a = |U_x(i+1,j,k) - U_x(i,j,k)|/\Delta x_b(i)$$
$$b = |U_y(i,j+1,k) - U_y(i,j,k)|/\Delta y_b(j)$$
$$c = |U_z(i,j,k+1) - U_z(i,j,k)|/\Delta z_b(k)$$
$$D = max(a,b,c)\alpha(i,j,k)/W(i,j,k)$$
$$E = max(D) \text{ over all zones}$$

The advection process also requires a limit on the time step in order that the interpolation of densities be accurate. For this criterion we form,

$$a = |V^x(i+1,j,k) - V_g^x(i,j,k)|/\Delta x_b(i)$$
$$b = |V^y(i,j+1,k) - V_g^y(i,j,k)|/\Delta y_b(j)$$
$$c = |V^z(i,j,k+1) - V_g^z(i,j,k)|/\Delta z_b(k)$$
$$F = max(a,b,c)\alpha(i,j,k)/W(i,j,k)$$
$$G = max(F) \text{ over all zones}$$

Finally, we choose Δt by,

$$\Delta t = \frac{K}{max(C, 4E, 2G)}$$

where K is a constant typically in the range from 0.25 to 0.50. The stability conditions are derived for linearized equations, hence we usually need K to be considerably smaller than unity. The viscosity and heat coefficients introduced for the Navier–Stokes equations are limited by the speed of sound and thus no further restriction is needed.

The difference equations as written are second-order accurate in time and, except for the artificial viscosity and the calculation of W, the artificial viscosity is not correctly time centered because it is formed with velocity differences which then change those velocities to the next time slice. One can calculate a new artificial stress with the new velocities and then use the average of the new and old stresses to obtain a better estimate of the viscous acceleration and energy deposition. To obtain good time centering for W would be difficult and has not been worked out yet.

If we are in doubt as to whether the errors in time centering are too large, we rerun the calculation with a reduced time step and check whether the results are significantly altered. Only calculations dominated by very strong shocks require reduced time steps.

We emphasize again that the state variables are also staggered in time in a manner such as to make the equations as nearly second order in time as possible. The following are all centered at the same time:

$$D, E, W$$

and the velocity-like quantities are entered $\Delta t/2$ away from D, E, W. Note that the W enters both into relating D and E to their proper densities ρ and $\rho\epsilon$ and into the relation of the velocities U_i and V^i. Restrictions on time step size may be required if W is changing too much per calculation cycle (see the time step discussion).

References

[1] L. Smarr, in *Sources of Gravitational Radiation*, ed. L. Smarr (Cambridge University Press, Cambridge, 1979), p. 245.

[2] L. Smarr, C. Taubes and J. R. Wilson, in *Essays in Relativity: A Festschrift for Abraham Taub*, ed. F. Tipler (Academic Press, New York, 1980), pp. 157–183.

[3] A. H. Taub, *Annu. Rev. Fluid Mech.*, **10**, 301 (1978).

[4] J. R. Wilson, *Astrophys. J.*, **173**, 431 (1972); **176**, 195 (1972).

[5] J. R. Wilson, in *Sources of Gravitational Radiation*, ed. L. Smarr (Cambridge University Press, Cambridge, 1979), p. 423.

4

Cosmological hydrodynamics

Most cosmology calculations utilize a homogeneous background spacetime in either a Friedmann–Robertson–Walker (FRW) metric or an equivalent extension. Calculations of local physical variables are treated as a perturbation on the background homogeneous expansion.

A natural question, however, is whether strong gravity waves in the early universe can cause a significant departure from this FRW treatment. There are also other cosmological questions which may require significant deviations from a FRW-plus-perturbations approach, such as the formation of primordial black holes, or the development of an inflating spacetime from inhomogeneous initial conditions. All such problems require the ability to solve for cosmological evolution in a metric more general than that of a simple perturbed FRW. Here we describe some attempts to model such cosmologies numerically.

4.1 Planar cosmology

In [1, 2, 4–7] a linear numerical cosmology program was developed to study how strong waves might affect the physics of the early universe. In particular, the paradigms for inflation, nucleosynthesis, and microwave anisotropy have been analyzed in this context, along with the question of whether strong gravity waves steepen in the early universe. That is, the nonlinear nature of general relativity could cause colliding waves to produce an even stronger superposition wave. In spherical symmetry such superpositions can even form black holes.

The simplest cosmology to study is that of planar symmetry. A system with planar symmetry can have gravity waves. In one-dimensional simulations, physical quantities are restrained to vary in one direction, say z. All fluid quantities are then functions of z and t only. However, gravity waves are transverse (like electromagnetic waves). Hence, the gravity waves are in the x, y plane.

To solve the field equations first select the metric,

$$g_{\mu\nu} = \begin{pmatrix} -\alpha + \beta_i \beta^i & \beta_x & 0 & \beta_z \\ \beta_x & \gamma_{xx} & 0 & 0 \\ 0 & 0 & \gamma_{yy} & 0 \\ \beta_z & 0 & 0 & \gamma_{zz} \end{pmatrix}, \tag{4.1}$$

where

$$\gamma_{ij} = A^2 \begin{pmatrix} 1 & 0 & 0 \\ 0 & h^2 & 0 \\ 0 & 0 & 1 \end{pmatrix}. \tag{4.2}$$

In the above, the *conformal factor* A is related to the scale factor R of the homogeneous Friedmann–Robertson–Walker metric ($h = 1$). The quantity h carries the anisotropy or metric shear. The extrinsic curvature is taken to be

$$K^i{}_j = \begin{pmatrix} K^x{}_x & 0 & K^x{}_z \\ 0 & K^y{}_y & 0 \\ K^z{}_x & 0 & K^z{}_z \end{pmatrix}, \tag{4.3}$$

with $-K^x{}_z = K^z{}_x$ since $\gamma_{xx} = \gamma_{zz}$. Next, we take

$$\mathrm{tr}(K) = K^x{}_x + K^y{}_y + K^z{}_z \equiv K \tag{4.4}$$

and define the *shear rate*

$$K_1 \equiv K^x{}_x - K^y{}_y. \tag{4.5}$$

Now K_1 and h carry the gravity wave degree of freedom. Note that this choice only allows one polarization. Two polarizations would require a more complicated representation of $K^i{}_j$ and γ_{ij}. Next, we need to select β_x and β_z so as to preserve the form of γ_{ij}.

As before the stress energy tensor is written

$$T_{\mu\nu} = \sigma U_\mu U_\nu + P g_{\mu\nu}, \tag{4.6}$$

where the inertial density σ is

$$\sigma = \rho(1 + \epsilon) + P. \tag{4.7}$$

And again, the time-like unit vector n_ν normal to each spatial slice is defined with the properties,

$$n_\nu = (-\alpha, 0, 0, 0), \qquad n^\nu = \left(\frac{1}{\alpha}, -\frac{\beta_i}{\alpha} \right), \qquad n^\nu n_\nu = -1.$$

Next, one can write for the Hamiltonian density (Eq. (1.33)),

$$\rho_H = n^\mu n^\nu T_{\mu\nu} = \rho(1 + \epsilon)W^2 + P(W^2 - 1), \tag{4.8}$$

and also the ADM momentum density,

$$s_i = S_i = -\sigma W^2 V_i = -\sigma W U_i, \tag{4.9}$$

where the covariant three-velocity as usual is

$$V_i \equiv U_i/W. \tag{4.10}$$

The spatial stress (Eq. (1.36)) is again just the space–space part of the energy momentum tensor,

$$S_{ij} = P\gamma_{ij} + \sigma W^2 V_i V_j = \sigma U_i U_j + \gamma_{ij} P.$$

The Hamiltonian constraint [28] is

$$R + K^2 - K^i{}_j K^j{}_i = 16\pi\rho_H \tag{4.11}$$

where R is the Ricci scalar of the three-space. From the evolution of the K_{ij}, the $\dot{K}$ equations can be written as

$$\gamma^{ij} D_i D_j \alpha = 4\pi\alpha\rho_\alpha + \beta^i D_i K - \dot{K}, \tag{4.12}$$

where

$$\rho_\alpha = \rho_H + S + \frac{1}{16\pi} K^j{}_i K^i{}_j. \tag{4.13}$$

In this one-dimensional cosmology, the lapse function can be evaluated from

$$\frac{\partial^2 \alpha}{\partial z^2} + \frac{\partial \alpha}{\partial z}\left(\frac{\partial \log A}{\partial z} + \frac{\partial \log h}{\partial z}\right) = 4\pi\alpha A^2\left[\frac{K^2}{16\pi} + \sigma(W^2 - 1/2) + P\right] - A^2 \dot{K}. \tag{4.14}$$

Next, the metric evolution equation

$$\dot{\gamma}_{ij} = -2\alpha\gamma_{ik}K^j{}_k + \gamma_{jk}D_i\beta^k + \gamma_{ik}D_j\beta^k \tag{4.15}$$

can be used to find conditions on $\vec{\beta}$. Imposing the symmetry condition $\dot{\gamma}_{xx} = \dot{\gamma}_{zz}$ gives

$$\frac{\partial\beta^z}{\partial z} = \frac{1}{2}\alpha\left(3K^z{}_z - K - K_1\right). \tag{4.16}$$

Then requiring $\gamma_{zx} = 0$ (or $\dot{\gamma}_{zx} = 0$) gives

$$\frac{\partial\beta^x}{\partial z} = 2\alpha K^z{}_x. \tag{4.17}$$

Similarly, $\gamma_{yz} = 0$ requires

$$\frac{\partial \beta^y}{\partial z} = 2\alpha K^y{}_z. \tag{4.18}$$

Here, the gauge choice is made to let $\beta^y = K^y{}_z = 0$.

Finally, the condition $\gamma_{xy} = 0$ requires

$$K^x{}_y = 0. \tag{4.19}$$

That the choice of $K^y{}_z = K^x{}_y = 0$ is acceptable must be demonstrated by the use of the $\dot{K}_{ij}$ equation for these components.

The Hamiltonian constraint in these coordinates is

$$\frac{\partial^2 h}{\partial z^2} + 2\frac{\partial h}{\partial z}\frac{\partial \log A}{\partial z} = h\left[-\frac{2}{A}\frac{\partial^2 A}{\partial z^2} + \left(\frac{\partial \log A}{\partial z}\right)^2\right.$$
$$\left. -\frac{A^2}{2}\left(\frac{K^j{}_i K^j{}_i - K^2}{32} + 8\pi\rho_H\right)\right]$$
$$-4\pi A^2 \rho_H + A^2\frac{(K^2 - K^j{}_i K^j{}_i)}{8}, \tag{4.20}$$

where ρ_H is given by Eq. (4.8).

Since the metric variable h is closely related to the gravity wave, it would be most desirable to to evolve h by an $\dot{h}$ equation and then solve the above equation for the metric conformal factor A. However, that equation is very nonlinear in A. Hence, it is easier to solve Eq. (4.20) for h and evolve A.

The evolution equation for A is

$$\dot{A} = \beta^z\frac{\partial A}{\partial z} - \frac{\alpha A}{2}\left(K + K_1 - K^z{}_z\right). \tag{4.21}$$

The momentum constraint

$$D_i\left(K^i{}_j - \gamma^i{}_j K\right) = 8\pi s_j \tag{4.22}$$

yields

$$\frac{\partial K^z{}_z}{\partial z} + \left(3K^z{}_z - K\right)\left(\frac{\partial \log A}{\partial z} + \frac{1}{2}\frac{\partial \log h}{\partial z}\right) + \frac{1}{2}K_1\frac{\partial \log h}{\partial x} = 8\pi S_z \tag{4.23}$$

and

$$\frac{\partial K^z{}_x}{\partial z} + \left(3\frac{\partial \log A}{\partial z} + \frac{\partial \log h}{\partial z}\right)K^z{}_x = 8\pi S_x. \tag{4.24}$$

These equations can be solved to find $K^z{}_z$ and $K^z{}_x$.

Finally, the evolution equation for K_1 is

$$\dot{K}_1 = \beta^z \frac{\partial K_1}{\partial z} + \alpha K_1 K - 2\alpha (K^z{}_x)^2$$
$$+ \frac{1}{A^2 h} \left[\frac{\partial \alpha}{\partial z} \frac{\partial h}{\partial z} + \alpha \left(\frac{\partial^2 h}{\partial z^2} + \frac{\partial h}{\partial x} \frac{\partial \log A}{\partial z} \right) \right]. \qquad (4.25)$$

This is a compléte set of equations, but one more equation,

$$\dot{h} = \alpha h K_1 + \beta^z \frac{\partial h}{\partial z}, \qquad (4.26)$$

is used to advance h to obtain a first guess for h before solving the Hamiltonian constraint equation.

4.1.1 Hydrodynamics for planar cosmology

As in previous chapters, the following hydrodynamic variables are defined:

$$W \equiv \alpha U^t, \qquad D = \rho W, \qquad E = \rho \epsilon W, \qquad S_i = \sigma W U_i, \qquad V^i = \frac{U^i}{U^t},$$

and the appropriate covariant derivatives,

$$\left(\rho U^\mu \right)_{;\mu} = 0, \qquad \left(T^\mu{}_\nu \right)_{;\mu} = 0, \qquad \left(U_\mu T^{\mu\nu} \right)_{;\mu} = 0,$$

are again used to derive the hydrodynamic equations of motion.
These become:
(1)   The continuity equation is written,

$$\dot{D} + D \frac{\dot{\gamma}}{\gamma} + \frac{1}{\gamma} \frac{\partial}{\partial z} (D \gamma V^z) = 0, \qquad (4.27)$$

where the first term on the left-hand side is the volume term and the second is the familiar transport term.
(2) The equations for the components of momentum density become

$$\dot{S}_x + S_x \frac{\dot{\gamma}}{\gamma} + \frac{1}{\gamma} \frac{\partial}{\partial z} (S_x \gamma V^z) = 0, \qquad (4.28)$$

and

$$\dot{S}_z + S_z \frac{\dot{\gamma}}{\gamma} + \frac{1}{\gamma} \frac{\partial}{\partial z} (S_z \gamma V^z) + \alpha \frac{\partial P}{\partial z} + \frac{1}{2} \frac{\partial}{\partial z} g^{\mu\nu} \frac{S_\mu S_\nu}{S^t} = 0, \qquad (4.29)$$

where the pressure force and gravitational forces appear only as the last two terms in the S_z equation. The gravitational acceleration can be further decomposed into

$$\frac{\partial}{\partial z} g^{\mu\nu} \frac{S_\mu S_\nu}{S^t} = 2\alpha \left[\sigma W^2 \frac{\partial \ln \alpha}{\partial z} - \sigma(W^2 - 1)\frac{\partial \ln A}{\partial z} - \left(S_x \frac{\partial \beta^x}{\partial z} + S_z \frac{\partial \beta^z}{\partial z} \right) \right].$$

(4.30)

Here, the first term on the right-hand side can be identified with the acceleration of observers at rest in a time slice, the second represents a velocity dependent gradient in the gravitational field, and the third term represents a kind of Coriolis-like force corresponding to the motion of the coordinates from one time slice to the next.

(3) The energy equation is now

$$\dot{E} + E\frac{\dot{\gamma}}{\gamma} + \frac{1}{\gamma}\frac{\partial}{\partial z}(E\gamma V^z) + \frac{P}{\gamma}\left[\frac{\partial}{\partial t}(W\gamma) + \frac{\partial}{\partial z}(W\gamma V^z) \right],$$

(4.31)

where the last two terms describe the relativistic analog of "PdV" work.

4.1.2 Solution of metric equations

One could choose $K(t)$ arbitrarily and put it into the equation for α and take whatever α comes out. It is better, however, to choose α to have some fixed mean value and then find $\dot{K}$ from the α equation such that the two are consistent. Another issue is that one must utilize a finite range for z for numerical purposes. A convenient choice is to fix the boundary conditions and then demand equality between the boundary values $z = 0$ and $z = L$ (where L is the length of the grid). This corresponds to a toroidal (or periodic) spatial geometry.

The equation for the lapse function is

$$\frac{\partial^2 \alpha}{\partial z^2} + \frac{\partial \alpha}{\partial z}\left(\frac{\partial \ln A}{\partial z} + \frac{\partial \ln h}{\partial z} \right) = \alpha A^2 \left[K^i{}_j K^j{}_i + \sigma(W^2 - \frac{1}{2}) + P \right] - A^2 \dot{K}.$$

(4.32)

To solve for α and $\dot{K}$, one first chooses $\alpha = 1$ and guesses a value for $\dot{K}$. The lapse equation is then integrated to find values of $\alpha(0)$ and $\alpha(L)$. If these two values are not equal a new guess for $\dot{K}$ is made and the integration repeated until $\alpha(0) = \alpha(L)$. This sets the value of $\dot{K}$. For flat space ($\alpha = 1$), the solution is

$$\dot{K} = K^i{}_j K^j{}_i + \sigma\left(W^2 - \frac{1}{2} \right) + P.$$

(4.33)

Now, $K^x{}_x = K^y{}_y = K^z{}_z = K/3$ so $K^i{}_j K^j{}_i = K^2/3$ and $W = 1$ in the absence of fluid velocity. The solution thus simplifies to:

$$\dot{K} = \frac{K^2}{3} + \frac{\sigma}{2} + P. \tag{4.34}$$

4.2 Applications

4.2.1 Nucleosynthesis

The planar cosmology defined above has been applied to the problem of inhomogeneous primordial nucleosynthesis [3, 6–8, 16–20, 24]. For this application one must add to the evolution equations the equations of nuclear abundance evolution [22, 25]

$$\frac{dY_i}{dt} = \sum_{ijk} N_i \left(\frac{Y_l^{N_l} Y_k^{N_k}}{N_l! N_k!} \langle n_k \sigma_{lk} v \rangle - \frac{Y_i^{N_i} Y_j^{N_j}}{N_i! N_k!} \langle n_j \sigma_{ij} v \rangle \right), \tag{4.35}$$

where the yield of each species Y_i is related to the mass fraction and the number density n_i, $Y_i = X_i/A = n_i/\rho N_A$, where A is the atomic mass of species i and N_A is Avogadro's number. The sum is over all reactions,

$$i + j \leftrightarrow k + l, \tag{4.36}$$

where σ_{ij} is the cross-section for the interaction of i with j to form k and l. The quantity N_i is the number of identically interacting species i; for example, in the $\alpha + \alpha \rightarrow {}^7\text{Li} + p$ reaction, $N_i = 2$, $N_j = 0$, $N_k = N_l = 1$. These factors prevent double counting of reacting species. The angled brackets denote an average over the Maxwellian distribution of velocities appropriate for the background temperature when the reactions are being evaluated.

In this application [8] perturbations and associated gravity waves were put in as initial conditions. The cosmological model was then evolved together with the nucleosynthesis equations under the constraint that the fluctuations in the microwave background temperature not exceed observed limits in $\Delta T/T$. Since the size of the horizon at the time of nucleosynthesis is only about 1 $M_\odot$, this is a relatively weak constraint since observational limits are at a much larger angular scale.

Figure 4.1 shows some representative results for the helium abundance in space and time. When averaged, these results show that the final primordial abundances do not vary much from the abundances derived from homogeneous models even though local abundances can vary by as much as a factor of two.

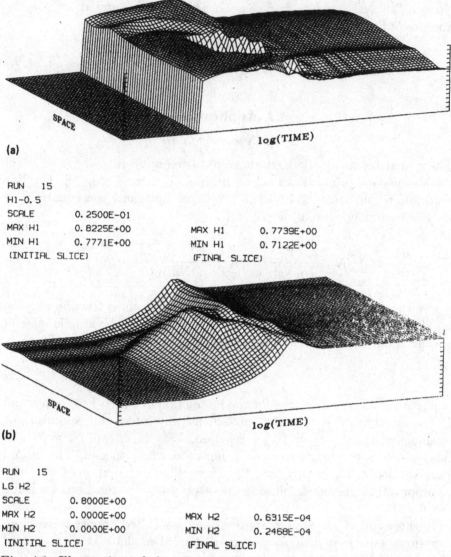

(a)

RUN   15
H1-0.5
SCALE   0.2500E-01
MAX H1   0.8225E+00   MAX H1   0.7739E+00
MIN H1   0.7771E+00   MIN H1   0.7122E+00
 (INITIAL SLICE)   (FINAL SLICE)

(b)

RUN   15
LG H2
SCALE   0.8000E+00
MAX H2   0.0000E+00   MAX H2   0.6315E-04
MIN H2   0.0000E+00   MIN H2   0.2468E-04
 (INITIAL SLICE)   (FINAL SLICE)

Fig. 4.1. Illustration of the abundances in spacetime computed in one of the inhomogeneous cosmologies considered in [8]. Shown are: (a) the proton mass fraction; (b) the deuterium mass fraction; (c) and (d) the ^{3}He mass fraction; and (e) the ^{4}He mass fraction. (Used by permision of *Nuclear Physics A*.)

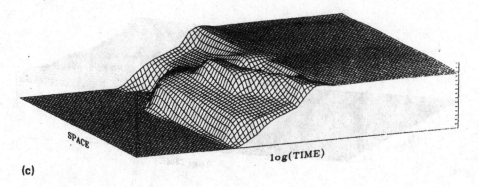

(c)

```
RUN    15
LG HE3
SCALE        0.8000E+00
MAX HE3      0.0000E+00        MAX HE3      0.2381E-04
MIN HE3      0.0000E+00        MIN HE3      0.1278E-04
(INITIAL SLICE)                (FINAL SLICE)
```

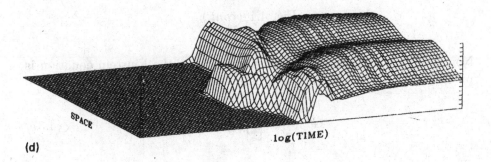

(d)

```
RUN    15
HE3
SCALE        0.2000E-05
MAX HE3      0.0000E+00        MAX HE3      0.2381E-04
MIN HE3      0.0000E+00        MIN HE3      0.1278E-04
(INITIAL SLICE)                (FINAL SLICE)
```

Fig. 4.1 (*cont.*)

4.2.2 Inflation

Another interesting application of the planar model discussed in this chapter has been to the effects of inhomogeneity on a developing inflationary [1, 11, 12, 21, 26] universe. To solve the fluid equations of motion in an inflationary universe one begins with the same constraints as discussed previously. In particular,

$$(T^{\mu\nu}_{fluid})_{;\nu} = 0.$$

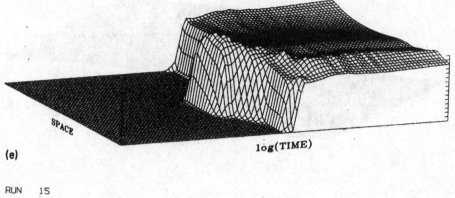

(e)

RUN   15
HE4
SCALE   0.2000E-01
MAX HE4   0.0000E+00
MIN HE4   0.0000E+00
(INITIAL SLICE)

MAX HE4   0.2893E+00
MIN HE4   0.2267E+00
(FINAL SLICE)

Fig. 4.1 (*cont.*)

Now, however, the energy momentum tensor in the Einstein equation is modified by the presence of the inflaton scalar field ϕ,

$$G^{\mu\nu} = 8\pi\left[T^{\mu\nu}_{fluid} + T^{\mu\nu}_\phi\right]. \tag{4.37}$$

The energy momentum tensor associated with the inflaton field is

$$T^{\mu\nu}_\phi = \partial^\mu\phi\partial^\nu\phi - g^{\mu\nu}\left[\frac{1}{2}\partial_\sigma\phi\partial^\sigma\phi + V(\phi)\right], \tag{4.38}$$

where the scalar field is taken to obey,

$$\phi^{;\mu}_{\;;\mu} = \Box\phi = \frac{dV(\phi)}{d\phi}, \tag{4.39}$$

where $V(\phi)$ is an effective potential associated with this scalar field. As in the $(3+1)$ ADM formalism (cf. Chapter 1) where a momentum conjugate to spatial curvature, K_{ij}, was introduced to obtain equations first order in time, one can introduce a momentum conjugate to the scalar field,

$$\pi = -\sqrt{-g}g^{t\nu}\partial_\nu\phi - \partial_i(\sqrt{-g}g^{ti}\phi). \tag{4.40}$$

Taking the covariant derivative of the inflaton field one obtains the following evolution equations:

$$\dot{\phi} = -\frac{1}{\sqrt{-g}} g^{tt} \left[\pi + \sqrt{-g} g^{ti} \partial_i \phi + \partial_i \left(\sqrt{-g} g^{ti} \phi \right) \right], \qquad (4.41)$$

$$\dot{\pi} = \partial_i \left[\sqrt{-g} g^{ij} \partial_j \phi + \phi \partial_t \left(\sqrt{-g} g^{it} \phi \right) \right] - \sqrt{g} \frac{\partial V(\phi)}{\partial \phi}. \qquad (4.42)$$

Now extra terms in the Einstein field equation arise since we have added $T_\phi^{\mu\nu}$ to $T_{fluid}^{\mu\nu}$. The solution of this system is straightforward but messy.

For the linear cosmology described in the previous section, the inflaton evolution equations are then,

$$\dot{\phi} = \frac{\alpha}{\gamma} + 2\beta^z \frac{\partial \phi}{\partial z} \left[\frac{1}{\gamma} \beta^z \frac{\partial \gamma}{\partial z} - \frac{\beta^z}{\alpha} \frac{\partial \alpha}{\partial z} + \frac{\partial \beta^z}{\partial z} \right] \phi, \qquad (4.43)$$

$$\dot{\pi} = \frac{\partial}{\partial z} \left[\left(\frac{1}{A^2} - \frac{\beta^{z\,2}}{\alpha^2} \right) \alpha \gamma \frac{\partial \phi}{\partial z} - \left(\frac{1}{\gamma} \beta^z \frac{\partial \gamma}{\partial t} - \frac{\gamma}{\alpha} \frac{\partial \beta^z}{\partial t} - \gamma \frac{\partial \beta^z}{\alpha^2} \frac{\partial \alpha}{\partial t} \right) \right]$$
$$- \alpha \gamma \frac{dV(\phi)}{d\phi}. \qquad (4.44)$$

4.2.3 Inflaton potential

Without question the largest uncertainty in the inflation paradigm is the nature and origin of the inflaton field effective potential. Indeed, many different forms and scenarios have been proposed (cf. [15, 23]) and although the microwave background (e.g. [16]) and large scale structure [13] provide some some constraint, the form of the potential and its spatial fluctuations are largely unknown. It is possible to show that a self-coupling higher than fourth order is not renormalizable (cf. [1]). Hence, a simple fourth-order polynomial is often assumed, although one should probably keep in mind that true inflaton potentials as derived from grand unified, super-symmetric, super-string theories, or quantum cosmology, are invariably much more complex (cf. [23]). Nevertheless, a common form is the Coleman–Weinberg potential,

$$V(\phi) = \lambda \left(\phi^2 - \sigma^2 \right)^2 \qquad (4.45)$$

or

$$V(\phi) = \lambda \left[\phi^4 \left(\log \frac{\phi^2}{\sigma} - \frac{1}{2} \right) + \frac{\sigma^4}{2} \right]. \qquad (4.46)$$

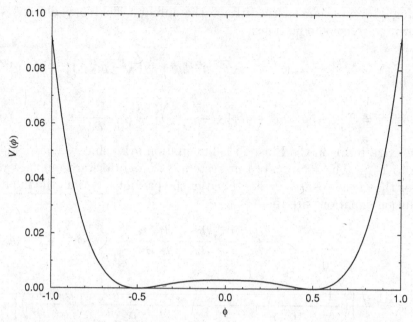

Fig. 4.2. Schematic illustration of a Coleman–Weinberg inflaton effective potential with $\sigma = 0.5$ and $\lambda = 0.1$.

This potential is sketched in Figure 4.2. The present limits [10] on the cosmological vacuum energy require that the present "true vacuum" ($\sigma = 0$) is effectively negligible compared to what it would have been during the inflation epoch,

$$V(\phi = \sigma) \approx \rho_\Lambda \approx \text{small}, \tag{4.47}$$

except perhaps in the scenario of cosmic quintessence (cf. [27]).

To study effects of initial inhomogeneities in the inflaton field Kurki-Suonio et al. [19] considered an initial condition, $\phi_0 = \sigma \cos kz$ at $t = 0$ so that V goes to zero at $\phi = \pm\sigma$ or $kz = n\pi$. The assumed initial fluctuation in the scalar field and effective potential are shown in Figure 4.3. The fluid equation of state for N relativistic particle species is just

$$\rho_{fluid} = \frac{N\pi^2}{50}T^4 \tag{4.48}$$

Figure 4.4 illustrates some results from [19]. Shown are the time evolution of the maximum and minimum of the scalar fields and the expansion factor R for $N = 2$ and different inflaton potentials labeled by λ and σ. For $\sigma = 0.5$ and $\lambda = 0.1$ there is no inflation, whereas for $\lambda \leq 0.01$ inflation occurs. On the other hand, for $\sigma = 0.05$ inflation requires λ ~ 10^{-5}. Otherwise, ϕ simply oscillates in time.

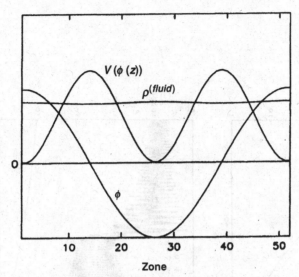

Fig. 4.3. Initial configuration of the scalar field ϕ adopted in [19]. (Used by permission of *The Physical Review*.)

In the models with no inflation there is a steep gradient in ϕ. Thus, $T_\phi^{zz} = \partial^2\phi/\partial z^2$ is large so that the energy in the ϕ field dominates. In models where inflation occurs, the amplitude of the oscillations is dampened by the expansion.

Such studies in planar cosmology address the question of whether initial matter anisotropies lead to a nonlinear growth of cosmological gravity waves. *A priori* it seems plausible that nonlinear gravitational effects could cause gravity waves to steepen. At least in planar cosmology the growth of gravity waves is almost linear and so they do not grow as the universe expands. Thus, such anisotropies are not important and primordial black holes do not form. In contrast, nonlinear effects are stronger in spherical geometry, hence we consider this symmetry next.

4.3 Spherical inhomogeneous cosmology

As the universe expands its contents change. That is, phase changes and gravitational clustering can occur. For example, above $T \sim 200$ MeV, quarks and gluons are present as well as photons and leptons ($\nu, \bar{\nu}, \mu, \bar{\mu}$, and e^+, e^- pairs). Below 100 MeV only bound hadrons (mostly pions and some nucleons) exist with the leptons. An interesting question, however, is the effect of phase transitions on the fluid pressure. In a first-order phase transition the universe cools to a coexistence temperature in which the two phases (in this example quark–gluon plasma and hadron gas)

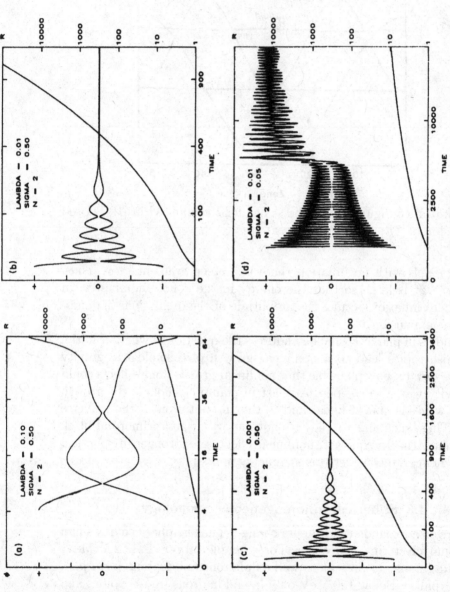

Fig. 4.4. Inhomogeneous inflation calculations for various Coleman–Weinberg potentials as labeled from [19]. The time axis is linear in $t^{1/2}$. The ordinate gives the ϕ scale with $+$ and $-$ denoting the $\pm\sigma$ true vacuum states: (a) shows an almost immediate domain formation; (b) and (c) are examples which produce inflation; (d) shows a phase transition without domain wall formation. (Used by permission of *The Physical Review*.)

are in pressure equilibrium [9]. In this coexistence phase the response of the pressure to changes in density is small, that is, the speed of sound $c_s^2 = \partial P/\partial \rho$ nearly vanishes. This is to be contrasted with a relativistic gas for which $\partial P/\partial \rho = 1/3$. If $\partial P/\partial \rho$ is small the fluid becomes unstable to collapse. Hence the prospect exists [14] for black hole formation during a cosmic phase transition, or any other epoch in which the pressure support is reduced.

The universe is most unstable during a given epoch on the scale of the event horizon. That scale corresponds to the largest mass that can self-interact. Nevertheless, this must be studied numerically since expansion and light travel times become significant.

4.3.1 The metric

In a spherical cosmology with inhomogeneous fluid flow, the metric is modified from the usual Robertson–Walker form into

$$g_{\mu\nu} = \begin{pmatrix} -\alpha + \beta_r \beta^r & \beta_r & 0 & 0 \\ \beta_r & \phi^4 & 0 & 0 \\ 0 & 0 & \phi^4 r^2 & 0 \\ 0 & 0 & 0 & \phi^4 r^2 \sin^2 \theta \end{pmatrix}. \tag{4.49}$$

where the conformal factor ϕ^4 relates to the scale factor. The shift β_r is selected so as to maintain the spatial metric isotropies. That is,

$$\gamma_{\theta\theta} = r^2 \gamma_{rr},$$

$$\gamma_{\phi\phi} = r^2 \sin^2 \theta \gamma_{rr},$$

$$\dot{\gamma}_{rr} = -2\alpha K_{rr} + 2D_r \beta_r,$$

$$\dot{\gamma}_{\theta\theta} = -2\alpha K_{\theta\theta} + 2D_\theta \beta_\theta.$$

If we assume $K^\theta{}_\theta = K^\phi{}_\phi$ and let $\mathrm{tr}(K^j{}_i) = K^r{}_r + K^\theta{}_\theta + K^\phi{}_\phi$, then the shift becomes

$$\beta^r = -\frac{r}{2} \int_r^\infty \alpha(3K^r{}_r - K)dr/r. \tag{4.50}$$

This conserves the isotropy of the metric.

The Hamiltonian constraint can now be used to find ϕ,

$$\frac{1}{r^2} \frac{\partial}{\partial r}\left(\frac{\partial \phi}{\partial r}\right) = -2\pi\phi^5 \Big[(\rho(1+\epsilon)W^2 + P(W^2 - 1)$$

$$+ \frac{1}{32\pi}\Big(3K_r^r - 2KK_r^r + K^2\Big)\Big]. \tag{4.51}$$

The lapse α is then deduced from

$$\frac{1}{r^2}\frac{\partial}{r}\left(r^2\frac{\partial}{\partial r}(\alpha\phi)\right) = 2\pi\alpha\phi^5\Big[\rho(1+\epsilon)(3W^2-1/2)+3P(W^2+1)$$

$$+\frac{7}{32\pi}\left(3K^r{}_r - 2KK^r{}_r + K^2\right)\Big]. \tag{4.52}$$

4.3.2 Quantum chromodynamics equation of state

The pressure and other thermodynamic variables can be determined [9] from the grand partition function and various derivatives of the thermodynamic potential, Ω. For a non-interacting massless quark–gluon plasma in quantum chromodynamics (QCD), for example, one can approximate Ω with,

$$\Omega_{qg} = -\frac{7\pi^2}{180}N_cN_fVT^4\left[1+\frac{30}{7\pi^2}\left(\frac{\mu_q}{T}\right)^2+\frac{15}{7\pi^4}\left(\frac{\mu_q}{T}\right)^4\right]-\frac{\pi^2}{45}N_gVT^4+BV, \tag{4.53}$$

where N_c is the number of colors, N_f is the number of quark flavors (2 or 3), $N_g = 8$ is the number of gluons, and B is the QCD vacuum energy.

4.3.3 Boundary conditions in spherical cosmology

To follow a finite spherical volume, an outer boundary condition is necessary. A good way to do this is as follows. First, one chooses a distribution of ρ and ϵ as initial conditions. Then, one makes approximate averages of $\bar\rho$ and $\bar\epsilon$ to define a standard Friedmann–Robertson–Walker (FRW) universe. From the FRM equations one can then find values for $\dot K(t)$, $\bar\rho(t)$, and $\bar\epsilon(t)$ at the boundary to use as boundary conditions for the Hamiltonian and lapse equations.

As boundary conditions on the lapse function $\alpha(R)$ (where R is the outer grid radius), one can take $\alpha(R) = 1$ and $(\partial\alpha/\partial r)_{r=R} = 0$. Similarly, the conformal factor ϕ is taken from the FRW solution and its derivative is also set to zero at the outer edge of the grid, $(\partial\phi/\partial r)_{r=R} = 0$. This insures that no acceleration occurs at the boundary.

In practice the initial distribution of $\rho(r)$ and $\epsilon(r)$ should be made with several zones at the outside with $\rho(r) = \bar\rho$ and $\epsilon(r) = \bar\epsilon$. This insures that the boundary will be well behaved (e.g. no fluid motion or heating at the boundary) as the system evolves. It is also usually necessary to iterate on the initial conditions by recalculating the conformal factor ϕ, then finding mean values for $\bar\rho$ and $\bar\epsilon$ near the boundary until self-consistency is achieved. The velocity inside the calculated sphere is also a free parameter.

The choice of the velocity distribution also affects the value of ϕ, so this also must be evaluated self-consistently.

Thus far, it has proven difficult to form primordial black holes using a simple Bag model QCD equation of state and small initial perturbations. However, it may be possible to form black holes if, for example, sterile neutrinos exist. The sterile neutrinos could provide a gravitational source but would not interact with the matter accelerations.

References

[1] R. Brandenberger, R. Kahn and W. H. Press, *Phys. Rev.*, **D28**, 1809 (1983); R. Brandenberger and R. Kahn, *Phys. Rev.*, **D29**, 2172 (1984); R. H. Brandenberger, *Rev. Mod. Phys.*, **57**, 1 (1985).

[2] J. Centrella, *Astrophys. J.*, **241**, 875 (1980).

[3] J. Centrella, in *Dynamical Spacetimes and Numerical Relativity*, ed. J. Centrella (Cambridge University Press, Cambridge, 1986), pp. 125–150.

[4] J. Centrella and R. A. Matzner, *Astrophys. J.*, **230**, 311 (1979).

[5] J. Centrella and R. A. Matzner, *Phys. Rev.*, **25**, 930 (1982).

[6] J. Centrella and J. R. Wilson, *Astrophys. J.*, **273**, 428 (1983).

[7] J. Centrella and J. R. Wilson, *Astrophys. J. Suppl. Ser.*, **54**, 229 (1984).

[8] J. Centrella, R. A. Matzner, T. Rothman and J. R. Wilson, *Nucl. Phys. B*, **266**, 171 (1986).

[9] G. M. Fuller, G. J. Mathews and C. R. Alcock, *Phys. Rev.*, **D37**, 1380 (1988).

[10] P. M. Garnavich, S. Jha, P. Challis, A. Clocchiatti, A. Diercks, F. V. Filippenko, R. L. Gilliland, C. J. Hogan, R. P. Kirshner, B. Leibundgut, M. M. Phillips, D. Reiss, A. G. Riess, B. P. Schmidt, R. A. Schommer, R. C. Smith, J. Spyromilio, C. Stubbs, N. B. Suntzeff, J. Tonry and S. M. Carroll, *Astrophys. J.*, **509**, 74 (1998).

[11] A. H. Guth, *Phys. Rev.*, **D23**, 347 (1981).

[12] A. H. Guth and S.-Y. Pi, *Phys. Rev. Lett.*, **49**, 1110 (1982); J. M. Bardeen, P. J. Steinhardt and M. S. Turner, *Phys. Rev.*, **D28**, 679 (1983).

[13] H. M. Hodges, G. Blumenthal, L. Kofman and J. Primack, *Nucl. Phys. B*, **335**, 197 (1990).

[14] K. Jedamzik and J. C. Niemeyer, *Phys. Rev.*, **59**, 124014 (2000).

[15] E. W. Kolb, *Phys. Rep.*, **227**, 5 (1993).

[16] H. Kurki-Suonio and G. J. Mathews, *Astrophys. J.*, **479**, 31 (1997).

[17] H. Kurki-Suonio and R. A. Matzner, *Phys. Rev.*, **D31**, 181 (1985).

[18] H. Kurki-Suonio and R. A. Matzner, *Phys. Rev.*, **D42**, 1047 (1990).

[19] H. Kurki-Suonio, J. Centrella, R. A. Matzner and J. R. Wilson, *Phys. Rev.*, **D35**, 435 (1987).

[20] H. Kurki-Suonio, R. A. Matzner, J. M. Centrella, T. Rothman and J. R. Wilson, *Phys. Rev.*, **D38**, 1091 (1988).

[21] A. Linde, *Phys. Lett.*, **129B**, 177 (1983).

[22] R. A. Malaney and G. J. Mathews, *Phys. Rep.*, **229**, 145 (1993).

[23] K. A. Olive, *Phys. Rep.*, **190**, 307 (1990).

[24] T. Rothman and R. A. Matzner, *Phys. Rev. Lett.*, **48**, 1565 (1982); *Astrophys. J.*, **257**, 450 (1982); **263**, 501 (1982); *Phys. Rev.*, **D30**, 1694 (1984).

[25] D. N. Schramm and R. V. Wagoner, *Annu. Rev. Nucl. Part. Sci.*, **27**, 37 (1977).

[26] P. J. Steinhardt, *Nature*, **345**, 47 (1990).

[27] M. Yahiro, G. J. Mathews, T. Kajino and K. Ichiki, *Phys. Rev.*, **D65**, 063502 (2002).

[28] J. W. York, Jr., in *Sources of Gravitational Radiation*, ed. L. Smarr (Cambridge University Press, Cambridge, 1979), p. 83.

5

Stellar collapse and supernovae

As discussed in Chapter 3, the treatment of the spherical collapse of a massive star to produce a supernova is sufficiently complex to warrant a separate chapter. It is believed that a Type II supernova arises from the delicate balance between energy deposited by escaping neutrinos from the core and the gravitational energy of collapsing outer layers. Thus, the relativistic energy and mass transport must be considered with high numerical accuracy to obtain a believable simulation. The model discussed herein includes the experience of about 30 years of development and should be of some guidance to those who wish to understand this fascinating phenomenon.

5.1 Collapse supernovae

A brief review of the scenario is as follows. Massive stars (i.e. 10 $\lesssim M/M_\odot \lesssim$ 30) evolve until the iron core exceeds $\sim$1–1.3 $M_\odot$. At this point there can be no more nuclear energy generation in the core. Neutrino emission, electron capture, and photodisintegration cool the inner $\sim$1 $M_\odot$ and remove pressure support from the core. The central density then rises. When the central density ρ_c approaches $\rho_c \sim 10^9$ g cm^{-3}, neutrino emission is so large that collapse becomes supersonic, i.e $v_{max} > c_s$, where c_s is the speed of sound. As the core collapses, the inner $\sim$0.7 $M_\odot$ collapses homologously (e.g. [2, 3, 9]). Once the core density exceeds nuclear density, $\rho_c > 2.5 \times 10^{14}$ g cm^{-3}, the pressure rises rapidly and collapse is halted. Matter continues to fall inward, however, so an outward moving shock wave is formed. This is referred to as the *core bounce*.

This shock expends most of its energy, however, into photodissociation. First nuclei in the remainder of the iron core are dissociated, and then elements above the core with lower atomic number. Behind the shock,

material has been disintegrated into nucleons and alpha particles and has
an entropy per baryon of $S/k \approx 10$. This disintegration of heavy nuclei
removes kinetic energy from the shock. If the shock breaks through with
enough remaining kinetic energy to unbind the outer parts of the star,
an explosion would occur by the "prompt" supernova mechanism (e.g.
[2, 9, 38]). In most calculations, however, the shock slows before it reaches
the outer boundary of the iron core and becomes an accretion shock, i.e.
moving outward in mass but not in radius.

A schematic illustration of the star and outer layers after ≈ 0.1 s is
shown in Figure 5.1. About this time we have a proto neutron star with
a radius of ~ 50 km surrounded by hot gas out to the shock wave at
~ 500 km. Beyond that matter is falling inward at roughly sonic speed.
The inner core is cooled by radiating energetic neutrinos which provide
an outlet for the gravitational binding energy released by the accreting
material.

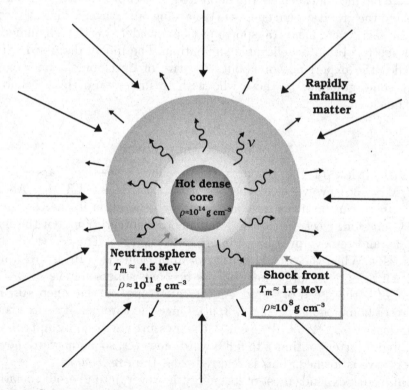

Fig. 5.1. Schematic illustration of the interior of a star undergoing a
supernova explosion at ≈ 0.1 s after core bounce.

In the hot gas region, material is cooling by neutrino emission and is being heated by neutrinos from the hot proto neutron star. The cooling rate is mostly due to charged current electron or positron capture on nucleons at the matter temperature T. This is given approximately [4] by

$$\text{cooling} = acT^4\sigma(T), \tag{5.1}$$

where $\sigma(T)$ is the average absorption cross-section for

$$p + e^- \rightarrow n + \nu_e$$
$$n + e^+ \rightarrow p + \bar{\nu}_e. \tag{5.2}$$

The reaction

$$e^- + e^+ \rightarrow \nu_e + \bar{\nu}_e \tag{5.3}$$

also makes a small contribution. Similarly, the heating from the absorption of energetic neutrinos escaping from from the neutrinosphere (radius of last neutrino scattering) is

$$\text{heating} = \frac{L}{4\pi r^2}\sigma(T_\nu) \tag{5.4}$$

where T_ν is the temperature of the neutrinosphere, and σ now refers to the absorption of neutrinos by the inverse of the above reactions as described in following sections.

One can estimate the net heating above the neutrinosphere. First, one can approximate the luminosity from the neutrinosphere as

$$L \approx \frac{acT_\nu^4}{4}4\pi R_\nu^2, \tag{5.5}$$

where R_ν is the neutrinosphere radius. Then, one can write

$$\sigma(T) \approx \sigma_0 T^2, \tag{5.6}$$

for the absorption cross-section. The rate of internal energy change due to neutrino absorption and emission then becomes

$$\dot{E} = \text{heating} - \text{cooling} = ac\sigma_0\left[\frac{T_\nu^6}{4}\left(\frac{R_\nu}{r}\right)^2 - T^6\right]. \tag{5.7}$$

The condition for net heating is thus

$$\frac{T_\nu}{T}\left(\frac{R_\nu}{r}\right)^{1/3} > 4^{1/6}. \tag{5.8}$$

This condition is satisfied in a region behind the shock and above the neutrinosphere.

As energy is deposited, the temperature rises and the entropy increases leading to the copious production of electron–positron pairs which can themselves scatter neutrinos resulting in yet more energy deposition. Material behind the shock expands away from or accretes back onto the newly formed proto neutron star creating an evacuated region. In this way, the entropy per baryon rises to S 200 after several seconds in the calculations of Mayle and Wilson [22, 23, 44], and continues to rise as the bubble is further evacuated. Although this region behind the shock is at low density, the entropy is so high that tremendous pressure builds up and revives the slowed shock's outward motion. If the revived shock breaks through the remainder of the initial core, a "delayed" supernova explosion occurs.

One interesting outcome of this picture is that the evacuated high entropy bubble provides an excellent site [29, 47] for the production [20] of heavy nuclei by rapid neutron capture (the *r-process*). A key component, however, is that the entropy must be high. In the model of Mayle and Wilson [22, 23, 44] the entropy per baryon reaches $S/k \approx 400$, while some approximate "wind" models (e.g. [32, 33]) lead to much lower entropy ($\sim$100).

Although this is a promising paradigm for the supernova mechanism, one must keep in mind that the binding energy of the neutron star is $\sim 3 \times 10^{53}$ erg while the kinetic energy of the resulting explosion is only $\sim 10^{51}$ erg. The process is very inefficient and therefore great care must be exercized in order to produce a believable result. In what follows we summarize the content of the model developed by Mayle and Wilson [22, 23, 44].

5.2 The physical model

5.2.1 The metric

Relativistic gravitational effects are not great for collapsing stars, but since the supernova process is so marginal they must be treated accurately. As alluded to in the introductory section, the spherical symmetry of this problem allows for a different formulation than the ADM metric described in Chapter 1. We describe here the Mayle and Wilson model. In this model general relativity is implemented by the use of Lagrangian hydrodynamics and a version of the May and White [21] metric:

$$ds^2 = -a^2 dt^2 + b^2 dm^2 + R^2(d\theta^2 + \sin^2\theta d\phi^2), \qquad (5.9)$$

where a Lagrangian rest mass coordinate m has been introduced. The metric coefficients a, b, and R are now functions of m (the rest mass) and time t. The angular coordinates are standard spherical coordinates on the unit sphere similar to Schwarzschild coordinates.

The state variables for the problem are now the radial position of a mass zone (R), the radial component of the fluid four-velocity (U), the matter internal energy (ϵ_M), the net charge per baryon (Y_e), and six energy distribution functions for the neutrinos $(F_{\nu_e}, F_{\bar{\nu}_e}, F_{\nu_\mu}, F_{\bar{\nu}_\mu}, F_{\nu_\tau}, F_{\bar{\nu}_\tau})$. In order to maintain good zoning, the state variables are remapped during each cycle onto a desired grid. Thus, the numerical scheme is a mixed Euler–Lagrange method as described in Section 2.3. However, this implementation differs from that described in Chapter 2. It does not use a grid velocity.

5.2.2 Energy momentum tensor

The next step is to define the energy momentum tensor relevant to this problem in these coordinates. Since neutrinos play such a crucial and distinct role from the rest of the matter, it is necessary to introduce a separation of internal energy density and pressure into individual contributions from matter and neutrinos. Hence, in obvious notation we write

$$\begin{aligned} \rho\epsilon &= \rho\epsilon_M + E_\nu \\ P &= P_M + P_\nu. \end{aligned} \tag{5.10}$$

An auxiliary pressure correction variable W_ν is also introduced:

$$W_\nu = \frac{(E_\nu - 3P_\nu)}{2}. \tag{5.11}$$

Since the neutrinos are not necessarily in thermodynamic equilibrium near the neutrinosphere, one must explicitly evolve distribution functions in neutrino energy and angle $F_i(E, \theta) = f_i(E, \theta)E^3$. The average energy E_ν, pressure P_ν, and neutrino flux Φ_ν are then defined by energy and angular integrations over this distribution function:

$$E_\nu = \sum_{i=1}^{6} \int F_i dE d\Omega_\nu, \tag{5.12}$$

$$\Phi_\nu = \sum_{i=1}^{6} \int F_i \cos(\theta) dE d\Omega_\nu, \tag{5.13}$$

$$P_\nu = \sum_{i=1}^{6} \int F_i \cos^2(\theta) dE d\Omega_\nu, \tag{5.14}$$

where $\Omega_\nu = 2\pi \sin\theta d\theta$ is the neutrino solid angle.

With these definitions, the energy momentum tensor becomes

$$
T^{\mu\nu} = \begin{pmatrix}
\frac{(\rho(1+\epsilon))}{a^\cdot} & \frac{\Phi_\nu 4\pi R^\cdot \rho}{a} & 0 & 0 \\
\frac{\Phi_\nu 4\pi R^\cdot \rho}{a} & P(4\pi R^2 \rho)^2 & 0 & 0 \\
0 & 0 & \frac{(P+W_\nu)}{R^\cdot} & 0 \\
0 & 0 & 0 & \frac{(P+W_\nu)}{R^\cdot \sin^\cdot \theta}
\end{pmatrix}. \tag{5.15}
$$

5.2.3 Evolution equations

As in previous chapters, having specified the metric and the energy momentum tensor it becomes possible to use the Einstein equation to find the metric coefficients.

We first introduce a new variable,

$$
\Gamma = \left(1 + U^2 - \frac{2M}{R}\right)^{1/2}, \tag{5.16}
$$

where M is the gravitational mass interior to R defined below, and $U \equiv U^R$ is the radial component of the four-velocity,

$$
U = \frac{1}{a}\frac{\partial R}{\partial t}. \tag{5.17}
$$

Obviously we have $U^\theta = U^\phi = 0$.

The quantity Γ (not to be confused with the equation of state index used in previous chapters) is related to $W = aU^t = (1 + U^2)^{1/2}$ of the (3+1) hydrodynamics formulation.

With this choice of variables, the metric can be rewritten as

$$
ds^2 = -a^2\left[1 - \left(\frac{U}{\Gamma}\right)^2\right]dt^2 - \frac{2aU}{\Gamma^2}dRdt + \frac{dR^2}{\Gamma^2} + R^2(d\theta^2 + \sin^2\theta d\phi^2), \tag{5.18}
$$

where the quantity $a \equiv 1/U^t$ is related to the gravitational red shift, and R has units of physical length. Proper distance is given by

$$
\text{proper distance} = \int \frac{dR}{\Gamma}. \tag{5.19}
$$

For the metric coefficient a the condition $T^{mj}{}_{;j} = 0$ implies

$$
a = \exp\left[\int_m^{m_{max}} \frac{dm}{\rho h}\left(\frac{\partial P}{\partial m} + \frac{bR^2\rho}{a}\frac{\partial}{\partial t}\left\{\frac{\Phi_\nu}{R^2\rho^2}\right\} - \frac{2b\Gamma}{R}W_\nu\right)\right], \tag{5.20}
$$

where m_{max} is the mass coordinate at the boundary of the numerical grid. The quantity $h \equiv 1 + \epsilon + P/\rho$ as defined in Chapter 3, and the metric variable, b, is given by

$$
b \equiv \frac{1}{4\pi R^2 \rho}. \tag{5.21}
$$

For the one-dimensional calculations, the metric coefficient $a(t, m)$ can be set to unity outside the grid and then the appropriate transformation used to find global time.

5.2.4 Matter equations

For the matter four-velocity acceleration we have,

$$\frac{1}{a}\frac{\partial U}{\partial t} = -\frac{\Gamma}{\rho h}\left[\frac{1}{b}\frac{\partial P}{\partial m} + \frac{R^2\rho^2}{a}\frac{\partial}{\partial t}\left\{\frac{\Phi_\nu}{R^2\rho^2}\right\} - \frac{2\Gamma}{R}W_\nu\right)\right]$$
$$-\frac{M}{R^2} - 4\pi RP. \tag{5.22}$$

The condition of baryon number conservation, $(\rho U^\mu)_{;\mu} = 0$, leads to auxiliary equations for the matter evolution:

$$\rho = \frac{\Gamma}{4\pi R^2}\left(\frac{\partial R}{\partial m}\right)^{-1}, \tag{5.23}$$

$$\frac{1}{a}\frac{\partial \rho}{\partial t} = -\rho\frac{1}{R^2}\frac{\partial}{\partial R}(R^2 U) + \frac{1}{2\Gamma}\rho R\Phi_\nu. \tag{5.24}$$

The gravitational mass is given by

$$M = 4\pi\int_0^m dm\frac{\partial R}{\partial m}R^2\left[\rho(1+\epsilon) + \frac{U}{\Gamma}\Phi_\nu\right], \tag{5.25}$$

$$\frac{1}{a}\frac{\partial M}{\partial t} = 4\pi R^2(UP + \Gamma\Phi_\nu), \tag{5.26}$$

$$\frac{1}{b}\frac{\partial M}{\partial m} = 4\pi R^2\left[\Gamma\rho(1+\epsilon) + U\Phi_\nu\right]. \tag{5.27}$$

The baryon rest mass of the star is simply given by the integral over the proper volume, $d(Vol) = 4\pi R^2 dR/\Gamma$,

$$M_0 = 4\pi\int R^2 dR\frac{\rho}{\Gamma}. \tag{5.28}$$

The matter internal energy evolves according to

$$\frac{1}{a}\frac{\partial \epsilon_M}{\partial t} = -P_M\frac{1}{a}\frac{\partial}{\partial t}\left(\frac{1}{\rho}\right) - \frac{1}{\rho}\sum_{i=1}^6\int \Lambda_i dE d\Omega_\nu, \tag{5.29}$$

where P_M is the matter pressure and the Λ_i are neutrino source terms to be discussed below.

The condition of lepton number conservation leads to an expression for the change in the average electron fraction (or charge per baryon) Y_e due to weak interactions,

$$\frac{\rho}{m_b}\frac{1}{a}\frac{\partial Y_e}{\partial t} = -\sum_i (\Lambda_i - \bar{\Lambda}_i)\frac{dq}{q}d\Omega_\nu, \tag{5.30}$$

where $q \equiv a\epsilon_\nu$, ϵ_ν is the neutrino energy, and q is the energy a neutrino would have if it were removed to infinity.

5.3 Numerical methods

5.3.1 Hydrodynamics

The hydrodynamic equations are solved using the Lagrangian formulation as outlined in Chapter 2 for the case of spherical symmetry and $V_g^i = V^i$. The acceleration equation (5.22), can be broken into two parts:

$$\frac{1}{a}\frac{\partial U}{\partial t}\bigg|_1 = -\frac{\Gamma}{\rho h}\left[\frac{1}{b}\frac{\partial P}{\partial m}\right] - \frac{M}{R^2}\left[1 + \frac{4\pi R^3 P}{M}\right], \tag{5.31}$$

and

$$\frac{1}{a}\frac{\partial U}{\partial t}\bigg|_2 = -\frac{\Gamma}{\rho h}\left[\frac{R^2\rho^2}{a}\frac{\partial}{\partial t}\left(\frac{\Phi_\nu}{R^2\rho^2}\right) - \frac{2\Gamma}{R}W_\nu\right]. \tag{5.32}$$

The latter acceleration is considerably smaller than the former. Therefore, these accelerations may be divided successfully. For example, the update of the four-velocity for (Eq. 5.31) now becomes:

$$U_i(t + dt) = U_i(t) - \frac{dt\Gamma}{\rho h}\left[\frac{1}{b}\frac{(P_i - P_{i-1})}{(m_i - m_{i-1})}\right] - \frac{M dt}{R^2}\left[1 + \frac{4\pi R^3 P}{M}\right], \tag{5.33}$$

with a similar update for Eq. (5.32).

The artificial viscosity acceleration is done in the same manner as was given in Chapter 2. The matter energy is advanced by the first term in Eq. (5.29) using $(\partial\rho/\partial t)$ from Eq. (5.24) and then the $(\rho R\Phi_\nu/2\Gamma)$ and W_ν terms are added in later during the neutrino diffusion calculation. At that time, the $(1/a)(\partial U/\partial t)|_2$ acceleration term is applied to the velocity. The red shift equation (5.20) is evaluated by combining the $\partial P/\partial m$ term from the acceleration calculation and the neutrino terms for $(1/a)(\partial U/\partial t)|_2$ which are formed during the neutrino diffusion calculation.

5.3.2 Time step

In the center of the developing proto neutron star the speed of sound is very high. Thus, the Courant condition (cf. Section 2.4.2) greatly restricts the time step. The central Courant time step is typically several microseconds. To run a collapse calculation for ~15 s then can be prohibitive. The neutrino sections of the calculation require much more computer time than the hydrodynamic calculations. Hence, after the explosion is underway it is possible to run up to approximately six hydrodynamic cycles before calling the neutrino diffusion calculation. An additional speed-up is possible by the fact that the Courant time step increases greatly going from inside the neutron star to outside. Therefore, after we compute a time step for every zone, we then divide the zones farther out in radius into shells such that all of the time steps in any shell are at least twice the minimum time step in the shell just interior to the shell under consideration. Each shell is assigned a time step such that the time steps increase by a factor of two from shell to shell. Then, the innermost shell is advanced by two cycles. The next shell is advanced one cycle so the inner two shells are at the same time. This is repeated and the third shell is then advanced one cycle. The process is repeated until all shells have advanced to the time of the outermost shell. Thus, for example, with six shells the overall time step is 2^5 times the innermost time step. Overall, by performing six hydrodynamic cycles per neutrino diffusion calculation, we have advanced in time by $32 \times 6 = 192$ times the innermost time step.

5.3.3 Remap

In the supernova computer program of Wilson and Mayle [44] the system is evolved for a time step in a completely Lagrangian sense (cf. Chapters 2 and 3). However, since matter is moving differently in different parts of the star, it is efficient to define a new spatial grid at each time step so as to represent best the physical structure of the star with a limited number of zones (typically 300). All of the physical matter and neutrino variables must then be *remapped* onto the new grid.

This process is similar to the transport algorithm given in Chapter 2. For example, the density remap is achieved by first obtaining a slope of density versus radius by the same equations as those leading to Eq. (2.107). Then an interface density q is found similarly to Eq. (2.108) except $V^i dt$ is replaced by the difference $[r_{new}(i) - r_{old}(i)]$. The matter to be exchanged between zones is

$$D_{exc} = \tilde{D} 4\pi (r_{new}^3 - r_{old}^3) \tilde{\Gamma}, \tag{5.34}$$

where $\tilde{\Gamma}$ is defined by Eq. (5.16) and interpolated to the center of the displaced volume $(r_{new}^3 - r_{old}^3)$. Finally, the new density is obtained,

$$D_{new}(i) = (D_{old}(i) \, Vol_{old}^p + (D_{exc}(i+1) - D_{exc}(i)) \, Vol_{new}^p, \tag{5.35}$$

where Vol^p is the proper zone volume for the chosen metric. It is

$$Vol^p(i) = (r_{i+1}^3 - r_i^3)\left(\frac{4\pi}{3\Gamma_i}\right).$$

(5.36)

As in Chapter 2, the exchanged mass D_{exc} is saved to use in the matter–energy exchange.

In the energy remap a boundary energy $\tilde{\epsilon}$ is found in the same manner as $\tilde{D}$ was determined above. Then, the energy exchanged is $\tilde{\epsilon} D_{exc}$ since ϵ is energy per unit mass.

Nuclear compositions are remapped similarly to ϵ. Neutrino distribution functions are remapped by the density algorithm described above. Variables such as the degree of decomposition of nuclei which are not state variables, but depend upon the equation of state, are found by using a quadratic interpolation to the new zone centers.

The velocity remap is performed in a manner similar to the momentum transport described in Chapter 2. A face-centered velocity is formed and the exchanged density and $\Gamma(\epsilon\rho)_{exc}$ are used to determine the exchanged momentum. If the change in kinetic energy is greater than a small fraction (10^{-10}) of the internal energy, the time step is reduced.

5.3.4 Opacity averaging

Most of the neutrino opacities vary as the square of the neutrino energy. The typical fractional zone width of the neutrino energy groups is $(2^{1/3} - 1)$. The Planck and Rosseland mean opacities are determined for each energy group and specific zone as follows. Below the photosphere, the neutrinos are assumed to be in a thermal distribution given by the local temperature and the appropriate averages are made. Above the photosphere, the neutrino distribution is assumed, for the purpose of group averaging, to have the same shape as that given by the temperature of the photosphere. The energy integrated absorption and transport cross-sections are also assumed to have this shape. The Planck opacity for emission is formed using the local temperature to determine the spectral shape.

5.4 Neutrino evolution equation

As seen in the previous section, a proper description of the matter evolution requires the determination of the neutrino distribution functions. This is a critical component of the supernova model.

In principle the distribution functions for the neutrinos are given from the relativistic Boltzmann equation [43, 48] for each neutrino type.

If we let $\mu \equiv \cos\theta$, then using the time–time and space–time components of the Einstein equation the relativistic Boltzmann equation for the neutrino distribution as a function of μ, q, R, and t can be written

$$
\frac{1}{a}\frac{\partial F_i}{\partial t} = \frac{\mu\Gamma}{aR^2}\frac{\partial}{\partial R}(aR^2 F_i) - \Gamma\left(\frac{1}{aR} - \frac{\partial}{R}\ln a\right)\left(\frac{\partial}{\partial\mu}(F_i(1-\mu^2))\right)
$$
$$
+ \frac{F_i}{a}\frac{\partial\ln\rho}{\partial R} + R\frac{\partial}{\partial R}\left(\frac{U}{R}\right)\left(\frac{\partial}{\partial\mu}\left[\mu(1-\mu^2)F_i\right] + \mu^2 q\frac{\partial F_i}{\partial q}\right)
$$
$$
+ \frac{q}{a}\frac{\partial F_i}{\partial q}\frac{\partial}{\partial t}\left(\ln\frac{R}{a}\right) + \kappa_i\rho(B - F_i)\left(1 + e^{q/aT}\right), \tag{5.37}
$$

where

$$
B \equiv g_i\left(\frac{q}{a}\right)^3 \frac{1}{e^{q/aT}+1}, \tag{5.38}
$$

with g_i the usual statistical factor. The quantity κ_i is the neutrino opacity.

5.4.1 Flux limited diffusion

The evaluation of this equation has been achieved in the past [43] by analyzing finite difference equations in the μ coordinate. However, for model computations with high grid resolution it is desirable to have a faster scheme without sacrificing accuracy. The most common solution is to introduce relativistic *flux limited diffusion*.

In this case the Boltzmann evolution is simplified to a diffusion equation of the form of Fick's law,

$$
\frac{1}{a}\frac{\partial G_i}{\partial t} \approx \nabla \cdot (D_i \nabla G_i). \tag{5.39}
$$

To achieve this simplification we begin by introducing angular moments of the distribution functions:

$$
G_i = \int F_i d\Omega_\nu, \tag{5.40}
$$

$$
H_i = \int F_i \cos(\theta) d\Omega_\nu, \tag{5.41}
$$

$$
K_i = \int F_i \cos^2(\theta) d\Omega_\nu. \tag{5.42}
$$

The angular integrated Boltzmann equation can then be written in terms of these moments,

$$
\frac{1}{a}\frac{\partial G_i}{\partial t} + q\frac{\partial G_i}{\partial q}\frac{1}{a}\frac{\partial}{\partial t}(\ln a) + \frac{1}{aR^2}\frac{1}{b}\frac{\partial}{\partial m}\left(aR^2 H_i\right) - \frac{U}{R}\left[q\frac{\partial}{\partial q}(G_i - 3K_i)\right]
$$
$$
- \frac{1}{a}\frac{\partial}{\partial t}(\ln(\rho))\left(G_i - q\frac{\partial}{\partial q}K_i\right) = \int \Lambda_i d\Omega_\nu. \tag{5.43}
$$

The summation in Eq. (5.43) is over the six neutrino and anti-neutrino types. The neutrino energy distribution functions G_i in Eq. (5.43) are functions of time, rest mass, and neutrino energy only. The angular degrees of freedom are integrated out in the right-hand side of Eq. (5.43). The source and sink terms, Λ_i, are complicated functions computed from all of the relevant neutrino–matter and neutrino–neutrino interactions [5, 22]. Some of these interactions we discuss below.

If a diffusion equation for neutrinos is to be used, closure relations (discussed below) are utilized to obtain expressions for the neutrino flux H_i and pressure K_i in terms of G_i. This closure process is also related to the ν–$\bar{\nu}$ annihilation rate in that the angular distribution of the neutrinos is crucial to both diffusion and annihilation. We have used the concept of *flux limitation* in terms of the quantity

$$x \equiv \frac{\lambda \Gamma |\partial G_i/\partial r|}{G_i},\tag{5.44}$$

where λ is the neutrino mean free path. When x is small ($x \ll 1$) the angular distribution is isotropic. In the limit of large x the distribution is free streaming. The complex problem of determining profiles of T, ρ, Y_e, G_i as functions of radius is described below in Section 5.4.5.

In brief, we write

$$H_i = D_i \Gamma \frac{\partial G_i}{\partial R}.\tag{5.45}$$

The method of flux limited diffusion consists of finding a form for D_i such that Eq. (5.39) remains valid from the diffusion limit where the neutrino mean free path is small compared to the case of neutrino free streaming over characteristic length scales of the simulation. This is achieved by defining the diffusion coefficient as follows:

$$D_i \approx \frac{\lambda_i}{3}\left(1 + h(x)x/3\right)^{-1}.\tag{5.46}$$

The quantity $h(x)$ is called the *flux limiter*. It is derived by constructing a Padé series which fits an exact beam calculation of neutrino flow from high to low density regimes in steady state,

$$h(x) \approx \frac{4 + \frac{x}{2} + \frac{x^{\cdot}}{8}}{1 + \frac{x}{2} + \frac{x^{\cdot}}{8}}.\tag{5.47}$$

Clearly, in the free streaming limit $x \gg 1$, $D \to G/\Gamma|\partial G \partial R|$, while in the short mean free path ($x \to 0$) limit, $D \to \lambda/3$ as it should. An explanation of how this form for $h(x)$ is derived is given in the next section.

5.4.2 Flux limiter beam calculation

One needs to have a good numerical understanding of the transport of neutrinos through the star for two purposes. One is to derive numerical values to which to fit the flux limiter $h(x)$ above. The other is to deduce a quantitative rate for neutrino annihilation as it transits the star.

The function $h(x)$ in Eq. (5.47) appears rather arbitrary. It serves to reduce the effective diffusion constant. Its form was selected by making calculations of the neutrino angular distribution and by taking the configuration at several times and assuming all the state variables (ρ, T, etc.) are fixed. Then the neutrino distribution is found by integrating the Boltzmann equation along many (several hundred) straight lines. That is, we ignore the gravitational bending of the neutrino trajectories. Then, with the detailed angular distribution determined, the neutrino flux can be calculated and compared to the neutrino flux calculated by the use of Eq. (5.45), (5.46), and (5.47). The agreement is usually good to a few percent with the $h(x)$ given in Eq. (5.47).

5.4.3 Neutrino annihilation beam calculation

The neutrino–anti-neutrino annihilation near and outside the photosphere is particularly dependent on the angular distribution $A(\mu)$. For the beam calculation, it is convenient to define a reduction factor $\tilde{R}$ for the neutrino annihilation rate relative to that of an isotropic neutrino distribution,

$$\tilde{R} = \frac{I}{JK}, \tag{5.48}$$

where

$$I \equiv \int \left(1 - \Omega_1 \cdot \Omega_2\right)^2 A(\mu_1)\bar{A}(\mu_2)d\mu_1 d\mu_2, \tag{5.49}$$

$$J \equiv \int A(\mu_1)\bar{A}(\mu_2)d\mu_1 d\mu_2, \tag{5.50}$$

$$K \equiv \int \left(1 - \Omega_1 \cdot \Omega_2\right)^2 d\mu_1 d\mu_2. \tag{5.51}$$

Following the method of [10, 16], Salmonson and Wilson [35] derived an asymptotic formula for the annihilation reduction factor which takes into account the gravitational bending of the neutrino trajectories. A reduction factor for annihilation is given by the following approximate analytic form:

$$\tilde{R} \approx 1 - 0.50f(x) - 0.50f(x)^2 + a(z), \tag{5.52}$$

where $f(x)$ is obtained from a fit to the beam calculation,

$$f(x) = (x/(3 + x(1 + 3e^{-x}))^2, \tag{5.53}$$

and $a(z)$ is given by the analytic solution derived in [35],

$$a(z) = (1 - z)^2(z^2 + 4z + 5)/8, \tag{5.54}$$

where the radial variable z is defined as

$$z = \sqrt{1 - (r_0/r)^2[(1 - 2m/r)/(1 - 2m/r_0)]}, \tag{5.55}$$

with r_0 the neutrinosphere radius. We note a word of caution that this model of Salmonson and Wilson assumes isotropic emission from a spherical surface. However, the photosphere region extends sufficiently far in radius that the assumption of isotropic emission is probably poor near the photosphere.

5.4.4 Neutrino pressure force

In the flux limiting approximation the matter acceleration (Eqs. (5.22), (5.31)) due to the neutrino pressure P_ν can be implemented by replacing the neutrino pressure with

$$P_\nu = \int \frac{D}{\lambda} G \frac{dq}{a}. \tag{5.56}$$

Note that in the free streaming limit $P_\nu \to 0$ as it should, while in the small λ regime, $P_\nu \to (1/3) \int G dq/a$.

In the short mean free path regimes, $W_\nu = (E_\nu - 3P_\nu)/2 \to 0$. Hence, $E_\nu \to \int G dq/a$. In the free streaming limit, $P_\nu = E_\nu$ while $W_\nu = -P_\nu$. These quantities will fall off as $1/R^2$ in the free streaming region as they should.

5.4.5 Neutrino angular distribution

We have reduced the problem of the angular distribution of neutrinos by making the "diffusion approximation". To obtain an energy integrated angular distribution function $A_i(R, \mu, t)$ for each neutrino species, we return to the Boltzmann equation (5.37) and integrate over the neutrino energy Q to obtain

$$\frac{1}{a}\frac{\partial A}{\partial t} = \frac{\mu\Gamma}{aR^2}\frac{\partial}{\partial R}(aR^2 A_i) - \Gamma\left(\frac{1}{aR} - \frac{\partial}{R}\ln a\right)\left(\frac{\partial}{\partial\mu}(A_i(1 - \mu^2))\right)$$
$$+ \frac{A_i}{a}\frac{\partial\ln\rho}{\partial R} + R\frac{\partial}{\partial R}\left(\frac{U}{R}\right)\left(\frac{\partial}{\partial\mu}\left[(\mu - \mu^2 - \mu^3)A_i\right]\right) +$$
$$- \frac{A_i}{a}\frac{\partial\ln(R/a)}{\partial t} + \Lambda_i \tag{5.57}$$

We take ρ, a, U as given functions of R and t, found by evolving the previous diffusion equation for G_i together with the hydrodynamic equations. Then taking $\partial A/\partial t = 0$, Eqs. (5.45), (5.46), and (5.47) can be solved by integrating along chords through the star to obtain the angular distribution. Similarly, we also integrate Eq. (5.37) itself along beams. In this way we check the accuracy of the form of Eq. (5.47) by comparing the neutrino fluxes and couplings. Evaluation of the neutrino interaction integrals Λ_i is an important part of this problem. We discuss these below in Section 5.5.

As another a check on this angular distribution, recall that the angular integrated first moment of the distribution should reduce to

$$H_i = \int \mu F_i d\Omega = -D_i \Gamma \frac{\partial G_i}{\partial R}. \tag{5.58}$$

On average we should have

$$\bar{D}_i = \frac{\int \mu A_i d\Omega}{\Gamma \partial (\int A_i d\Omega)/\partial R}. \tag{5.59}$$

A comparison of $\bar{D}_i$ evaluated this way with that evaluated from the $D_i(x)$ formula shows the consistency of this approach and is a good code check in numerical calculations.

5.4.6 Operator splitting for the neutrino distribution

The most difficult equations to solve during the supernova evolution are those determining the neutrino distribution functions G_i. To achieve this, an operator splitting of Eq. (5.43) is used, whereby

$$\frac{1}{a} \frac{\partial G_i}{\partial t} = \left(\hat{O}_{red} + \hat{O}_{diff} + \hat{O}_{hydro} + \hat{O}_{source} + \hat{O}_{sink} \right) G_i, \tag{5.60}$$

where the $\hat{O}$ are differential or integral operators which are labeled according to a physical meaning (e.g. red = red shift) attributed to each term in Eq. (5.43):

$$\hat{O}_{red} G_i = -\frac{1}{a} \frac{\partial}{\partial t} (\ln(a)) q \frac{\partial}{\partial q} G_i, \tag{5.61}$$

$$\hat{O}_{diff} G_i = -\frac{1}{aR^2} \frac{1}{b} \frac{\partial}{\partial m} \left(aR^2 H_i \right), \tag{5.62}$$

$$\hat{O}_{hydro} G_i = \frac{U}{R} \left[q \frac{\partial}{\partial q} (G_i - 3K_i) \right] + \frac{1}{a} \frac{\partial}{\partial t} (\ln (\rho)) \left(G_i - q \frac{\partial}{\partial q} K_i \right) q \frac{\partial}{\partial q} G_i, \tag{5.63}$$

$$\left(\hat{O}_{source} + \hat{O}_{sink}\right)G_i = \int \Lambda_i d\Omega_\nu. \tag{5.64}$$

The neutrino diffusion equation (cf. Eq. (5.43)) is solved in two steps. During the main hydrodynamic calculation the $\dot{a}$, $\dot{\rho}$, and $(U/R)[\partial(G_i - 3K_i)/\partial q]$ are evaluated and used to advance G_i. Then in the neutrino diffusion section a relatively simple diffusion equation,

$$\frac{1}{a}\frac{\partial G_i}{\partial t} + \frac{1}{aR^2}\frac{1}{b}\frac{\partial}{\partial m}\left(aR^2 H_i\right) - \frac{U}{R}\left[q\frac{\partial}{\partial q}(G_i - 3K_i)\right] = \int \Lambda_i d\Omega_\nu, \tag{5.65}$$

is left to be solved.

5.5 Neutrino–matter interactions

We now consider in detail the principal neutrino source and sink terms $\int \Lambda_i d\Omega_\nu$ for the neutrino Boltzmann equation (5.43), and Eq. (5.64). It has become clear over time that the many complex neutrino interactions are ultimately responsible for the success or failure of the supernova mechanism. These can be categorized as follows:

Charged current interactions:

Electron capture

$$\nu_e + n \rightleftharpoons e^- + p,$$

$$\nu_e + A(Z, N) \rightleftharpoons e^- + A(Z+1, N-1).$$

Electron scattering

$$\nu_e + e^- \rightleftharpoons e^- + \nu_e,$$

$$\bar{\nu}_e + e^- \rightleftharpoons e^- + \bar{\nu}_e.$$

Annihilation

$$\nu_e + \bar{\nu}_e \rightleftharpoons e^- + e^+.$$

Neutral current interactions:

Scattering

$$\nu_e + e^- \rightleftharpoons e^- + \nu_e,$$

$$\nu_e + A(Z, N) \rightleftharpoons \nu_e + A(Z, N),$$

$$\nu_e + p \rightleftharpoons \nu_e + p,$$

$$\nu_e + n \rightleftharpoons \nu_e + n.$$

Annihilation

$$\nu_e + \bar{\nu}_e \rightleftharpoons e^- + e^-.$$

There are comparable reactions for the μ and τ neutrinos. One must also consider inelastic and coherent processes. We now describe the numerical implementation of some of these processes in more detail.

5.5.1 Electron capture

We consider first the change of the electron-neutrino distribution due to electron capture by heavy nuclei.

$$e^- + A(Z, N) \rightleftharpoons \mu_e + A(Z - 1, N + 1), \tag{5.66}$$

where $A(Z, N)$ represents a heavy nucleus with Z protons and N neutrons. Electron capture of this form is only considered in regions below nuclear matter density in which the nuclei can be treated in *nuclear statistical equilibrium (NSE)*. The relation between the electron energy ϵ_e and the neutrino energy ϵ_ν and the reaction threshold Q_r is simply

$$\epsilon_e = \epsilon_\nu + Q_r, \tag{5.67}$$

where the reaction thresholds are approximated by

$$Q_r \approx \hat{\mu} + \frac{\epsilon^*}{1 + (kT/\epsilon^*)^2}, \tag{5.68}$$

where $\hat{\mu}$ is difference in chemical potential between the neutrons and protons, and $\epsilon^* \approx 3$ MeV is a mean excitation energy. At low temperatures and densities $Q_r \approx \epsilon^*$, while at high temperatures and densities

$$Q_r \approx \hat{\mu}. \tag{5.69}$$

The cross-section for electron (or neutrino) capture is motivated by the calculations of Fuller *et al.* [14, 15]. The capture is approximated with the free nucleon cross-section, but with only a fraction of the protons and neutrons allowed to participate. The number $\tilde{N}$ of protons or neutrons that can participate for neutron number $N > 40$ is approximated as

$$\tilde{N} = 2.512 \times 10^{-3/T_{MeV}}, \tag{5.70}$$

for neutron numbers $N \leq 40$, $\tilde{N} = 2.512$. This condition on neutron number represents the *thermal Pauli blocking* of the electron capture reactions as described in [14, 15].

The cross-sections for free proton electron capture σ_{ep} can be written [39]

$$\sigma_{ep} = \left(\frac{1 + 3\alpha_S^2}{8}\right) \sigma_0 \left(\frac{\epsilon_\nu}{m_e}\right)^2. \tag{5.71}$$

Similarly, the cross-section for neutrino capture on a neutron is

$$\sigma_{\nu_e n} = \left(\frac{1 + 3\alpha_S^2}{4}\right) \sigma_0 \left(\frac{\epsilon_e}{m_e}\right)^2, \tag{5.72}$$

where $\sigma_0 = 16\pi G_F(m_e h/c)^2 = 1.7 \times 10^{-44}$ cm^2, with G_F the weak interaction coupling constant, and m_e the electron rest mass. The quantity $\alpha_S \approx 1.2$ adds a strong interaction effect.

The electron capture source and sink terms can now be written

$$\int \Lambda_{EC} d\Omega_\nu = \left(\frac{\rho X_A}{\langle A\rangle m_B}\right)\tilde{N}c\epsilon_\nu\left[\sigma_{ep}n_e(1-f_{nu_e}) - \sigma_{\nu_e n}\left(\frac{F_{\nu_e}}{\epsilon_\nu}\right)(1-f_e)\right], \tag{5.73}$$

where c is the relative velocity of the collision (taken here to be the speed of light). The factor $(1 - f_e)$ is the blocking factor for an electron Fermi–Dirac distribution f_e. Similarly, the $(1 - f_{\nu_e})$ factor represents neutrino blocking with f_{ν_e} determined from

$$F_{\nu_e} \equiv \frac{4\pi}{(hc)^3}\epsilon_\nu^3 f_{\nu_e}. \tag{5.74}$$

Dividing F_{ν_e} by ϵ_ν as on the right-hand side of Eq. (5.73) gives the number density of neutrinos in the energy interval between ϵ_ν and $\epsilon_\nu + d\epsilon_\nu$. The quantity n_e is the number density of electrons in the interval between ϵ_e and $\epsilon_e + d\epsilon_e = \epsilon_\nu + d\epsilon_\nu + Q_r$. The expression for n_e is

$$n_e = \frac{8\pi}{(hc)^3}f_e\epsilon_e^2\sqrt{1-(m_e/\epsilon_e)^2}. \tag{5.75}$$

Using these relations we have the following evolution equation for the electron neutrino flux:

$$\frac{1}{a}\frac{\partial F_{\nu_e}}{\partial t} = c\rho\kappa_{EC}(B - F_{\nu_e}), \tag{5.76}$$

where the opacity is

$$\kappa_{EC} = \frac{X_A}{\langle A\rangle m_B}\tilde{N}\left[\frac{1+3\alpha_s^2}{4}\right]\sigma_0\left(\frac{\epsilon_\nu + Q_r}{m_e}\right)^2, \tag{5.77}$$

and B is the equilibrium distribution

$$B = \frac{4\pi}{(hc)^3}f_e\epsilon_e^3 = \frac{4\pi}{(hc)^3}\frac{\epsilon_e^3}{\exp\left[(\epsilon_e - \mu_e)/kT\right] + 1}. \tag{5.78}$$

As in Bowers and Wilson [5] we have used the following approximate formula for the electron chemical potential μ_e:

$$\mu_e = \frac{\mu_{0e}}{1 + (2.20kT/\mu_{0e})^2}, \tag{5.79}$$

where

$$\mu_{0e} = hc\left(\frac{3}{8\pi}\frac{\rho Y_e}{m_B}\right)^{1/3}. \tag{5.80}$$

Here, μ_{0e} is the $T = 0$ limit for μ_e.

As the electron–neutrino distribution is evolved, one must also update the matter internal energy ϵ_M and the net charge per baryon Y_e. For these we can similarly write

$$\frac{1}{a}\frac{\partial \epsilon_M}{\partial t} = \int c\kappa_{EC}(B - F_{\nu_e})d\epsilon_\nu, \tag{5.81}$$

and

$$\frac{1}{a}\frac{\partial Y_e}{\partial t} = \int c\kappa_{EC}(B - F_{\nu_e})\frac{d\epsilon_\nu}{\epsilon_\nu}. \tag{5.82}$$

5.5.2 Neutrino–electron elastic scattering

Neutrino elastic scattering, e.g.

$$\nu_e + e^- \rightarrow \nu_e + e^- \tag{5.83}$$

alters the neutrino spectra and exchanges energy with the electrons. The process is complicated by the nonthermal nature of the neutrino distributions throughout much of the core. Details of the treatment are given in Bowers and Wilson [5] which we now summarize here.

As the density in the core decreases to ρ 10^9 g cm^{-3}, the neutrino mean free path is typically much larger than the size of the core and the neutrinos escape. These neutrinos will be emitted from the core with an energy ($\sim$15 MeV) typical of their last scattering. There will be, however, an occasional scattering between these neutrinos and electrons in the low density regions.

In the low density regions the electrons are relativistic and may be degenerate or nondegenerate. Combining the expressions for the scattering cross-sections from Tubbs and Schramm [40] into analytic expressions valid in both the degenerate and nondegenerate limits, one can write the energy and angle averaged cross-section as

$$\sigma_{e,i} = \frac{3}{8}\sigma_0 c_{e,i}\left(kT + \frac{1}{4}\mu_e\right)\frac{\epsilon}{(m_e c^2)^2}, \tag{5.84}$$

where $i = \nu_e, \bar{\nu}_e, \nu_\mu, \bar{\nu}_\mu, \nu_\tau, \bar{\nu}_\tau$, and μ_e is the electron chemical potential. The constant factors $c_{e,i}$ are given in terms of the Weinberg angle Θ_W as:

$$c_{e,\nu_e} = (1 + 2\sin^2(\Theta_W))^2 + \frac{4}{3}\sin^4(\Theta_W), \tag{5.85}$$

$$c_{e,\bar{\nu}_e} = \frac{1}{3}(1 + 2\sin^2(\Theta_W))^2 + 4\sin^4(\Theta_W), \tag{5.86}$$

$$c_{e,\nu_\mu} = (1 - 2\sin^2(\Theta_W))^2 + \frac{4}{3}\sin^4(\Theta_W), \tag{5.87}$$

$$c_{e,\bar{\nu}_\mu} = \frac{1}{3}(1 - 2\sin^2(\Theta_W))^2 + 4\sin^4(\Theta_W), \tag{5.88}$$

$$c_{e,\nu_\tau} = (1 - 2\sin^2(\Theta_W))^2 + \frac{4}{3}\sin^4(\Theta_W), \tag{5.89}$$

$$c_{e,\bar{\nu}_\tau} = \frac{1}{3}(1 - 2\sin^2(\Theta_W))^2 + 4\sin^4(\Theta_W). \tag{5.90}$$

The mean neutrino energy loss per collision is

$$\left(\frac{\Delta\epsilon}{\epsilon}\right)_{coll} \approx \frac{1}{2}(1 - 4kT/\epsilon). \tag{5.91}$$

In the low density regions, the up-scattering of neutrinos is ignored. Therefore, neutrinos are only scattered if their energy exceeds, $\epsilon_{min} > 4kT$. The change in the spectrum due to the down-scattering of neutrinos is described by the following difference equation:

$$F'_{k,j} = F_{k,j} + n_e c\Delta t\left[\sigma_{j+1}\tilde{F}_{k,j+1}\frac{\Delta\epsilon_{coll,j+1}}{\Delta\epsilon_{j+1}} - \sigma_j\tilde{F}_{k,j}\frac{\Delta\epsilon_{coll,j}}{\Delta\epsilon_{j1}}\right], \tag{5.92}$$

where

$$\tilde{F}_{k,j} = F_{k,j}/[1 + n_e c\Delta t\sigma_j(\Delta\epsilon_{coll}/\Delta\epsilon_j)]. \tag{5.93}$$

Here, the factor of neutrino energy loss per collision $\Delta\epsilon_{coll}$ is required to produce the correct heating. This factor follows from the requirement

$$-\Delta Q_\nu = \sum_j (F'_{k,j} - F_{k,j})\Delta\nu_j$$

$$= \sum_{j>j_{min}} (n_e c\sigma_j\Delta t)\left(\tilde{F}_{k,j}\frac{\Delta\epsilon_{coll,j+1}}{\Delta\epsilon_{j+1}}\right)\Delta\epsilon_{coll,j}. \tag{5.94}$$

The total matter heating from elastic scattering of all neutrinos is then

$$\rho\frac{d\epsilon_M}{dt} = \frac{d}{dt}(Q_{\nu_e} + Q_{\bar{\nu}_e} + Q_{\nu_\mu} + Q_{\bar{\nu}_\mu} + Q_{\nu_\tau} + Q_{\bar{\nu}_\tau}). \tag{5.95}$$

5.5.3 Intermediate density scattering

At high densities, neutrinos are trapped and well represented as Fermi–Dirac distribution functions. However, for intermediate matter densities, $3 \times 10^9 < \rho < 10^{12}$, the neutrinos can interact with the matter but are not trapped. In this transition region one must explicitly evolve the neutrino distributions. It is possible to treat scattering in this regime by solving the full Boltzmann equation (e.g. [27, 43, 48]). A good approximation, however, [45] is to treat the neutrino scattering as an energy diffusion problem in the Fokker–Planck approximation,

$$\frac{dF}{dt} = \epsilon \frac{\partial}{\partial \epsilon} \left\{ K \left[F(1 - F/\alpha \epsilon^3) + kT \left(\frac{\partial F}{\partial \epsilon} - \frac{3F}{\epsilon} \right) \right] \right\}, \tag{5.96}$$

where the diffusion coefficient K is related to the relaxation time τ_c and the neutrino energy exchange per collision,

$$K = \left| \frac{\Delta \epsilon}{\epsilon} \right|_{coll} \tau_c^{-1}. \tag{5.97}$$

The fractional energy loss per collision (5.91) in this case is adjusted for the appropriate energy threshold [5–7, 45].

The relaxation time is taken as

$$\tau_c^{-1} = \frac{n_e c \sigma_{e,\nu}}{1 + 2 n_e c \sigma_{e,\nu} \Delta t |\Delta \epsilon/\epsilon|_{coll}}, \tag{5.98}$$

where the electron neutrino scattering cross-section in Weinberg–Salam theory is given by

$$\sigma_{e,\nu} = c_{e,\nu} \sigma_H, \tag{5.99}$$

with

$$\sigma_H = \begin{cases} 7.66 \sigma_0 \epsilon T & : T \geq \mu_e, \\ 0.98 \sigma_0 \epsilon \mu_e & : T < \mu_e \text{ and } \epsilon > \mu_e, \\ 1.48 \sigma_0 \epsilon \frac{T}{\mu_e} (1 + 11.6 T/\epsilon)(1 + 0.259 \epsilon^2/T) & : T < \mu_e \text{ and } \epsilon < \mu_e, \end{cases} \tag{5.100}$$

where $\sigma_0 = 4 G^2 m_e^2 \hbar^2 / \pi c^2 = 1.7 \times 10^{-44}$ cm^2, and ϵ is in MeV.

Now, the finite difference Fokker–Planck equation becomes

$$F'_{k,j} = F_{k,j} + \frac{\epsilon_j}{\Delta \epsilon_j} \left\{ K_{j+1/2} \left[T \frac{F'_{k,j+1} - F'_{k,j}}{\Delta \epsilon_{j+1/2}} - \frac{3T}{2} \left(\frac{F'_{k,j+1}}{\epsilon_{j+1}} + \frac{F'_{k,j}}{\epsilon_j} \right) \right. \right.$$
$$\left. + F'_{k,j+1/2} Q_{k,j+1/2} \right] - K_{j-1/2} \left[T \frac{F'_{k,j} - F'_{k,j-1}}{\Delta \epsilon_{j-1/2}} \right.$$
$$\left. \left. - \frac{3T}{2} \left(\frac{F'_{k,j}}{\epsilon_j} + \frac{F'_{k,j-1}}{\epsilon_{j-1}} \right) + F'_{k,j-1/2} Q_{k,j-1/2} \right] \right\}, \tag{5.101}$$

where

$$Q_{j,j+1/2} = 1 - \frac{F_{k,j+1}}{2\alpha \epsilon_{j+1}^3} - \frac{F_{k,j+1}}{2\alpha \epsilon_{j+1}^3}, \tag{5.102}$$

and $K_{j+1/2}$ refers to the proper centering of the diffusion coefficient, Eq. (5.97).

This difference equation can be written more compactly as

$$F'_{k,j} - F_{k,j} = \frac{\epsilon_j}{\Delta \epsilon_j}(F_{j+1} - F_{j-1/2}),$$ (5.103)

where F_{j+1} is defined as $K_{j+1/2}$ times the first term in brackets in Eq. (5.101). Neutrino number is then conserved exactly as long as

$$\sum_{j=1}^{J}(F'_{k,j} - F_{k,j})\frac{\Delta \epsilon_j}{\epsilon_j}\sum_{j=1}^{J}(F_{j+1} - F_{j-1}) = F_{J+1/2} - F_{-1/2} = 0.$$ (5.104)

One can then define $F_{k,J+1}$ such that $F_{J+1/2} = F_{-1/2}$ [7].

The coefficients $F_{k,j\pm1}$ are defined at the energy nodes. However, they can be written as a linear combination of group centered values:

$$F_{k,j+1/2} = a_j F_{k,j+1} + (1 - a_j)F_{j,k},$$ (5.105)

where the coefficients a_j can be determined [7] by requiring that a thermal distribution be achieved in the steady state solution to Eq. (5.101). A simpler approach, however, is to require that the steady state solution be given by

$$F + T\left(\frac{\partial F}{\partial \epsilon} - \frac{3F}{\epsilon}\right) = 0.$$ (5.106)

When finite differenced, this leads to a solution for the coefficients,

$$a_j = \frac{\frac{3T}{2}\left[\left(\frac{F_{k,j\cdot\cdot}}{\epsilon_{j\cdot\cdot}}\right) + \left(\frac{F_{k,j}}{\epsilon_j}\right)\right] - T\left[\frac{(F_{k,j\cdot\cdot} - F_{k,j})}{\Delta\epsilon_{j\cdot\cdot/\cdot}}\right] - F_{k,j}}{F_{k,j+1} - F_{k,j}}.$$ (5.107)

Inserting this into Eq. (5.101), the finite differenced Fokker–Planck equation then assumes the form

$$A_j + B_j F'_{k,j-1} + C_j F'_{k,j} + D_j F'_{k,j+1} = 0,$$ (5.108)

which is the implicit solution of a diffusion equation.

5.5.4 Neutrino annihilation

A major heating mechanism at late times outside the proto neutron star is neutrino pair annihilation,

$$\nu_e + \bar{\nu}_e \rightleftharpoons e^- + e^+.$$ (5.109)

The cross-section for neutrino interactions is proportional to the neutrino energy squared in the rest frame of the reaction,

$$\sigma_{eff} = \sigma_0 \int \left(1 - \Omega_1 \cdot \Omega_2\right)^2 A(\mu_1)\bar{A}(\mu_2)d\mu_1 d\mu_2.$$ (5.110)

Relativistic corrections to the neutrino annihilation rate can be important. See [35] and Section 5.4.3 for a discussion of how to implement them.

5.5.5 Pair production of neutrinos in the core

An important mechanism for neutrino production in the core is by pair annihilation,

$$e^+ + e^- \leftrightarrow \nu_i + \bar{\nu}_i. \tag{5.111}$$

There are also contributions from plasmon decay,

$$\gamma_{pl} \leftrightarrow \nu_i + \bar{\nu}_i, \tag{5.112}$$

and by the weak conversion of energetic electrons and positrons into μ or τ mesons:

$$e^- \leftrightarrow \nu_e + \bar{\nu}_\mu + \mu^-, \tag{5.113}$$

$$e^+ \leftrightarrow \bar{\nu}_e + \nu_\mu + \mu^+, \tag{5.114}$$

with a similar conversion for τ mesons.

The μ and τ neutrinos are expected to behave similarly. They are described by distribution functions $H(\nu_i)$. In general, H and $\bar{H}$ are not Fermi–Dirac distributions and must be evolved in time.

An excellent approximation [5] to the spectral evolution due to pair annihilation is given by

$$\left(\frac{dH}{dt}\right)_{pair} = g\frac{\dot{E}_{pair}}{48kT^9}\frac{\epsilon^4 T^4}{e^{\epsilon/T}+1}\left[1 - \frac{H}{g\alpha\epsilon^3}\left(\frac{e^{\epsilon/T}}{\beta T^4}\int\frac{Hd\epsilon}{g\alpha}+1\right)\right], \tag{5.115}$$

where $\beta = 5.682$ is a normalization constant, and $\dot{E}_{pair}$ is taken to be the contribution from both thermal pair annihilation and plasma decay,

$$\dot{E}_{pair} = \dot{E}_{th} + \dot{E}_{plas}. \tag{5.116}$$

The thermal rate of pair annihilations is given by

$$\dot{E}_{th} = \int\int(\epsilon_e^- + \epsilon_e^+)\sigma_L v dn_e^+ dn_e^-, \tag{5.117}$$

where σ_L is the weak interaction cross-section. The following simple analytic expression provides a reasonable approximation in both the degenerate and nondegenerate relativistic limits [5]:

$$\dot{E}_{th} \approx \frac{69.3\sigma_0}{4\pi m_e^2}\left[(C_v - 1)^2 + C_A\right](kT)^9 f_T(x), \tag{5.118}$$

where T is in MeV, and

$$f_T(x) = \frac{1}{1 + 0.07e^x} + 0.007X^5 e^{-x}, \tag{5.119}$$

with $x = \mu_e/kT$. The energy rate from plasmon decay can be reproduced sufficiently accurately with

$$\dot{E}_{plas} = 2.85 \times 10^{15} (C_v - 1)^2 T^6 Y^6 e^{-Y_f} f_p(Y) \text{ erg cm}^{-3} \text{ s}^{-1}, \tag{5.120}$$

where

$$Y = \frac{\hbar\omega_p}{kT} = \frac{0.288}{T} \times 10^{-2} \rho^{1/3}, \tag{5.121}$$

and

$$f_p = \frac{15(1 + Y^2)}{15 + Y^2}. \tag{5.122}$$

5.5.6 Neutrino–nucleus interactions

Coherent scattering of neutrinos from nuclei includes corrections for the nuclear finite size and ion–ion correlation effects [5, 40]. After angular integration, the coherent scattering cross-section can be written

$$\sigma_{A,Coh} = \frac{2\,(\bar{A} - X_B)}{3} \frac{\sigma_0 \nu^2}{4} \frac{1}{(m_e c^2)^2} \left[\frac{1}{2}(1 - 2\sin^2\theta_W)(2Z - 1) - \sin^2\theta_W\right]^2, \tag{5.123}$$

where $\bar{A}$ is the atomic weight averaged over all heavy nuclei and helium

$$\bar{A} = X_A A + 4X_{He} + X_B. \tag{5.124}$$

Nuclear structure effects are added by multiplying $\sigma_{A,Coh}$ by a factor [40]

$$F(\epsilon) = \exp\left[-\frac{1}{2}\frac{r_A \epsilon}{\hbar c}\right], \tag{5.125}$$

where $r_A \approx 10^{-13}$ cm is the nuclear radius. Ion–ion correlation effects are similarly approximated by multiplying $\sigma_{A,Coh}$ by a factor $S(X)$ given in [5]:

$$S(X) = \begin{cases} \frac{6}{bX_*} \left[1 - \frac{X}{2} - \frac{X_*}{6} + \frac{1-X}{X} \ln(1 - X)\right] & : X \leq X_1, \\[2ex] \frac{6}{bX_*} \left\{\frac{X_*}{3} - (X - 1)\left[\ln(1 - X_1) + X_1 + \frac{X_*}{2}\right]\right\} & : X > X_1, \\ \quad + 1 - 3(X_1/X)^2 + 2(X_1/X)^3 \end{cases} \tag{5.126}$$

where

$$X = \frac{b}{2\Gamma} \frac{4(h\nu)^2 a^2}{(\hbar c)^2}, \qquad X_1 = b/(1+b), \qquad b = 3\Gamma/10 - 1, \qquad (5.127)$$

and

$$\Gamma = \frac{(ZAe)^2}{akT}, \qquad (5.128)$$

with the interior nuclear separation given by $a = (3/4\pi n_A)^{1/3}$.

Neutrinos can also scatter inelastically from nuclei and eject a nucleon, leaving the daughter nucleus in an excited state. For example,

$$\nu_i + Z^A \rightarrow \nu_i' + Z^{(A-1)*} + n. \qquad (5.129)$$

To include this effect one can take

$$\dot{F}_i = f_H(\epsilon_\nu - \epsilon_\nu^0)^2 \frac{c\sigma_0 \rho N_A}{(m_e c^2)^2} = H, \qquad (5.130)$$

where the neutrino–nucleon interaction cross-section is $\sigma_0 \approx 1.7 \times 10^{-44}$ cm^{-2} and ϵ_ν^0 is the threshold energy for neutrino induced nucleon emission, typically 15–18 MeV. The numerical implementation of Eq. (5.130) is to update the neutrino distributions F_i' with

$$F_i' - F_i = \frac{F_- + f_H H dt}{1 - H dt}. \qquad (5.131)$$

The energy deposited by the neutrino is then

$$\epsilon_i' = \epsilon_i + (F_i' - F_i)k(\epsilon_\nu = \epsilon_\nu^0)/\rho, \qquad (5.132)$$

where k is the Boltzmann constant.

5.6 Equation of state

Another key component of the supernova problem is the equation of state. The components of matter which contribute significantly to the matter equation of state during various epochs of the supernova collapse include photons, electrons, positrons, pions, free neutrons, protons, and atomic nuclei. Since material is optically thick to photons, we include photons with matter particles in the equation of state.

In the simulations of Wilson and Mayle [44] matter is assumed to be in local thermal equilibrium (one temperature in a zone) but not necessarily in chemical equilibrium (i.e. the weak reactions have not necessarily equilibrated). The independent variables chosen for their equation of state are the temperature T, the matter rest mass density ρ, and the net charge per baryon Y_e. Nuclear isotopic abundances are also followed. If the temperature and densities are such that nuclear statistical equilibrium can be maintained, then the abundances themselves are simple functions of the state variables as follows.

5.6.1 Nuclear statistical equilibrium (NSE)

In NSE the individual elemental abundances are given by the nuclear Saha equation,

$$X(Z,A) = \frac{g(Z,A)}{2^A}\left[\frac{m_\mu \rho^{A-1}}{m_n^{A-Z} m_p^Z}\right]A_\mu^{5/2}\left(\frac{2\pi\hbar^2}{m_\mu kT}\right)^{3(A-1)/2} X_n^{A-Z} X_p^Z$$

$$\times \exp\left\{\frac{B(Z,A)}{kT}\right\}, \tag{5.133}$$

where $g(Z,A)$ is the nuclear partition function,

$$g(Z,A) = \sum_i E_i(2J_i + 1)\exp\left\{E_i/kT\right\}, \tag{5.134}$$

m_μ is the atomic mass unit, while m_p and m_n are the proton and neutron rest masses, respectively. The quantity $B(Z < A)$ is the nuclear binding energy:

$$B(Z,A) = (A-Z)m_n + Zm_p - A_\mu(Z,A)m_\mu, \tag{5.135}$$

where $A_\mu(Z,A)$ is the atomic mass, $A_\mu = M(Z,A)/m_\mu$.

The neutron and proton mass fractions are determined from the electron fraction Y_e by means of the constraints:

$$\sum_i \left(\frac{Z}{A}\right) X_i = Y_e, \tag{5.136}$$

and

$$\sum_i X_i = 1. \tag{5.137}$$

In the model of Mayle and Wilson [22], baryonic matter in NSE is represented as free neutrons, free protons, helium, and one representative heavy nucleus

$$\langle A \rangle = \sum_{i>4} A_i X_i \tag{5.138}$$

5.6.2 Nuclear burning

When matter is not in nuclear statistical equilibrium (NSE), the individual isotopic abundances are independent variables which must be evolved dynamically. In this case, the nuclear matter evolution is approximated by a nine element nuclear burn network (i.e. n, p, ^{4}He, ^{12}C, ^{16}O, ^{20}Ne, ^{24}Mg, ^{28}Si, ^{56}Ni) as described in [46]. Thus, we must write the matter pressure P_M and matter internal energy per gram ϵ_M as functions of composition as well, e.g. $P_M(\rho, T, Y_e, X_i)$ where the X_i are elemental mass fractions.

5.6.3 Photons, electrons, positrons, and pions

The photon contribution to the matter pressure P_γ and the photon energy density E_γ is given by the usual Stefan–Boltzmann law,

$$E_\gamma = 3P_\gamma = aT^4, \tag{5.139}$$

where $a = 1.37 \times 10^{26}$ erg cm^{-3} MeV^{-4}. Numerical integrations over Fermi–Dirac distributions are utilized to obtain tables of electron plus positron pressure and energy density. The pions are assumed to obey a dispersion relation of the form

$$\epsilon_\pi^2 = m_\pi^2 + p_\pi^2 R\left(1 + \frac{\Lambda^2\chi}{1 - g'\Lambda^2\chi}\right). \tag{5.140}$$

The quantity in brackets is a many body term which takes into account the polarization of the nuclear medium. It is given by [12]

$$\Lambda^2\chi = -\frac{4.53\omega\rho}{m_\pi^2(\omega^2 - \epsilon^2)} \exp\left[-2\left(\frac{p}{7m_\pi}\right)^2\right], \tag{5.141}$$

with

$$\omega = \sqrt{m_\Delta^2 + p^2} - m_N. \tag{5.142}$$

The pion contribution to pressure and energy density is then evaluated by applying Eq. (5.140) to the relevant Bose–Einstein integrals [24, 25]. See Chapter 2 and Section 5.6.6 for analyses and implementation of the above pion model.

5.6.4 Baryons

The equation of state for baryonic particles is separated into three regions: (1) matter below nuclear density but not in NSE; (2) matter below nuclear density and in NSE; (3) matter above nuclear density. The equation of state for each of these regions is constructed from the Helmholtz free energy per baryon

$$F = -kT \ln Z, \tag{5.143}$$

where Z is the relevant partition function. The pressure and energy density then follow from the usual relations of statistical mechanics,

$$P_M = \frac{\rho^2}{m_B} \left(\frac{\partial F}{\partial \rho} \right)_{(T,X_i,Y_e)}, \tag{5.144}$$

$$\epsilon_M = -\frac{T^2}{m_B} \left(\frac{\partial F}{\partial T} \right)_{(T,X_i,Y_e)}. \tag{5.145}$$

5.6.5 Baryons below nuclear density not in NSE

In this region, the free energy per baryon is taken as the sum of the ideal gas free energy F_g plus a Coulomb correction F_C. For the ideal gas component one can write

$$F_g = \sum_i \left[\frac{kT}{A_i} \ln \left(\frac{X_i \rho \alpha}{T^{3/2} A_i^{5/2}} \right) \right]. \tag{5.146}$$

The thermal wavelength per electron charge α is written

$$\alpha = \frac{h^3}{3(2\pi k m_B)^{3/2}}, \tag{5.147}$$

where i runs over the nuclei in the burn network. The Coulomb term is

$$F_C = -\frac{1}{3} \left(\frac{\rho}{m_B} \right)^{1/3} e^2 \langle A \rangle^{2/3} Y_e^2. \tag{5.148}$$

With these relations the baryonic pressure and energy per gram become

$$P_M = kT \frac{\rho}{m_B} \left(\sum_i \frac{X_i}{A_i} \right) - \frac{1}{9m_B} \left(\frac{\rho}{m_B} \right)^{4/3} e^2 \langle A \rangle^{2/3} Y_e^2, \tag{5.149}$$

and

$$\epsilon_M = \frac{3}{2} \frac{kT}{m_B} \left(\sum_i \frac{X_i}{A_i} \right) - \frac{1}{3m_B} \left(\frac{\rho}{m_B} \right)^{1/3} e^2 \langle A \rangle^{2/3} Y_e^2. \tag{5.150}$$

5.6.6 Baryons below nuclear density and in NSE

In this case for purposes of generating the equation of state, the nuclear constituents are represented simply as free nucleons, alpha particles, and a representative heavy average nucleus in nuclear statistical equilibrium. The free energy per baryon is then written as a sum of constituents,

$$F = F_n + F_p + F_\alpha + F_A. \tag{5.151}$$

The various free energies become

$$F_p = X_B Y_p \left\{ \epsilon_{p0} W + \epsilon_N (1 - W) \right.$$
$$\left. + \frac{3}{2} kT \left[\sqrt{1 + \zeta_p^2} - \ln \left(\frac{1 + \sqrt{1 + \zeta_p^2}}{\beta \zeta_p} \right) \right] \right\}, \tag{5.152}$$

$$F_n = X_B Y_n \left\{ \epsilon_{n0} W + \epsilon_N (1 - W) \right.$$
$$\left. + \frac{3}{2} kT \left[\sqrt{1 + \zeta_n^2} - \ln \left(\frac{1 + \sqrt{1 + \zeta_n^2}}{\beta \zeta_n} \right) \right] \right\}, \tag{5.153}$$

$$F_\alpha = X_\alpha \left\{ \epsilon_{\alpha 0} W + \epsilon_N (1 - W) + \frac{kT}{4} \left(\frac{X_\alpha \rho \alpha}{T^{3/2} 4^{5/2}} \right) \right\}, \tag{5.154}$$

$$F_A = X_A \left[-\frac{1}{3} \left(\frac{\rho}{m_B} \right)^{1/3} e^2 \langle A \rangle^{2/3} Y_A^2 + S_E (Y_{Fe} - Y_A)^2 \right.$$
$$\left. + \epsilon_N (1 - W) + \frac{3}{4} \rho^{4/3} Y_A^2 b(Y_e) + \frac{kT}{A} \left(\frac{X_A \rho \alpha}{g_A T^{3/2} A^{5/2}} \right) \right], \tag{5.155}$$

where
$$\langle A \rangle = 194.0 (1 - Y_e)^2 (1 + X + 2X^2 + 3X^3), \tag{5.156}$$

where
$$X \equiv \left(\frac{\rho}{7.6 \times 10^{13} (\text{ g cm}^{-3})} \right)^{1/3}. \tag{5.157}$$

The quantity A in the above is taken as $A = \langle A \rangle$ for $\langle A \rangle < 100$ and $A = 100$ for $\langle A \rangle \geq 100$. The quantity W in the above is a weighting factor which interpolates between low density and high density regimes.

The transition from subnuclear density to supranuclear density is expected to be smooth. The reason is that progressively larger nuclei are formed more or less continuously as the density increases. When a relativistic Thomas–Fermi representation of the electrons is evaluated at subnuclear density, the electron energy is lowered by more than ~ 1 MeV. The electrostatic nuclear energy increases in magnitude as well, and the transformation of nuclei from spheres to other shapes (spaghetti, lasagna, etc.) also lowers the energy of the medium by about 1 MeV. The net result is that the pressure and energy are very smooth functions of density near the nuclear saturation density. The weighting factor,

$$W = \left(1 - \frac{\rho}{\rho_N}\right)^2, \tag{5.158}$$

is chosen to represent this smooth transition.

Normal ^{56}Fe is taken as the zero of binding energy. The energy required to dissociate ^{56}Fe into free nucleons is $\epsilon_{p0} = 8.37$ MeV for protons, while for neutrons it is $\epsilon_{n0} = 9.15$ MeV. The quantity ϵ_N is the binding energy per nucleon at nuclear density. It is defined in the next subsection.

The quantity ρ_N is the density at which nuclear matter becomes a uniform sea of nucleons. This was found by fitting the saturation density of nuclear matter ($P_M(\rho, T = 0, Y_e) = 0$) for ρ as a function of Y_e. The zero temperature result was chosen to simplify the problem of making a smooth transition between the three equation of state regimes. The result is

$$\rho_N = (2.66 \times 10^{14})(1 - (1 - 2Y_e)^{5/2}) \text{ g cm}^{-3}. \tag{5.159}$$

The quantities ζ_n and ζ_p are a measure of the degeneracy of the free baryons. They are defined by

$$\zeta_n = \frac{B(\rho Y_n X_B)^{2/3}}{kT}, \qquad \zeta_p = \frac{B(\rho Y_p X_B)^{2/3}}{kT}, \tag{5.160}$$

where the constant B is chosen to be

$$B = \frac{3}{10}\left(\frac{3}{8\pi}\right)^{2/3} \frac{h^2}{m_B^{5/3}} \tag{5.161}$$

such that $B(\rho Y_i X_B)^{2/3}$ is the energy per baryon of a zero temperature nonrelativistic fermion ideal gas.

In the above, X_B is the free baryon mass fraction while X_α and X_A are the mass fractions of ^{4}He and heavy nuclei in obvious notation. The quantities Y_p and Y_n are the fractions of free nucleons in protons or neutrons, respectively. Clearly, $Y_p + Y_n = 1$. The quantity Y_A is the average Z/A for heavy nuclei.

The constant β appearing in Eqs. (5.152) and (5.153) is determined such that the translational parts of F_p and F_n reduce to the correct nondegenerate limit ($T \to \infty$, $\zeta_i \to 0$). That is,

$$\frac{3}{2}kT\left[\sqrt{1+\zeta_n^2} - \ln\left(\frac{1+\sqrt{1+\zeta_n^2}}{\beta\zeta_n}\right)\right] \to kT\ln\left(\frac{x_B\rho Y_i\alpha}{sT^{3/2}}\right). \tag{5.162}$$

This requirement implies

$$\beta = \left(\frac{\alpha}{2}\right)^{2/3}\left(\frac{3}{eB}\right) = 0.781. \tag{5.163}$$

The expression for $b(Y_e)$ in Eq. (5.155) is determined by the condition that the Coulomb contribution to the pressure at $\rho = \rho_N$ is canceled by the term proportional to $b(Y_e)$. This gives,

$$b(Y_e) = \frac{1}{18}\left(\frac{1}{m_B}\right)^{1/3} e^2 \langle A \rangle^{2/3}\left[\frac{1}{\rho_N} + 2\left(\frac{\partial ln\langle A \rangle}{\partial \rho}\right)_{\rho=\rho_N}\right]. \tag{5.164}$$

The expression for the statistical weight of the heavy nucleus g_A appearing in Eq. (5.155) is

$$\frac{1}{A}\ln g_A = \frac{3}{2}\left\{\left[1 - \sqrt{1 + \left(\frac{T}{T_S}\right)^2}\right]\frac{T}{T_S} + \ln\left[\frac{T}{T_S} + \sqrt{1 + \left(\frac{T}{T_S}\right)}\right]\right\}, \tag{5.165}$$

where

$$T_S = 8(\text{ MeV})\left(1 + 2\frac{\rho}{\rho_N}\right). \tag{5.166}$$

The constant $S_E = 120$ MeV is derived for a symmetry energy of 30 MeV per nucleon. The constant $Y_{Fe} = 0.464$ is the fraction of protons in ^{56}Fe.

The chemical potentials are found from the free energy as

$$\mu_n = \left(\frac{\partial F}{\partial X_B} - \frac{Y_p}{X_B}\frac{\partial F}{\partial Y_p}\right), \tag{5.167}$$

$$\mu_p = \left(\frac{\partial F}{\partial X_B} + \frac{Y_n}{X_B}\frac{\partial F}{\partial Y_p}\right), \tag{5.168}$$

$$\mu_\alpha = 4\left(\frac{\partial F}{\partial X_\alpha}\right), \tag{5.169}$$

$$\mu_{nA} = \left(\frac{\partial F}{\partial X_A} - \frac{Y_A}{X_A}\frac{\partial F}{\partial Y_A}\right), \tag{5.170}$$

$$\mu_{pA} = \left(\frac{\partial F}{\partial X_A} + \frac{(1 - Y_A)}{X_A}\frac{\partial F}{\partial Y_A}\right), \tag{5.171}$$

where μ_p, μ_n, and μ_α are the chemical potentials of free protons, neutrons, and alpha particles. The quantities μ_{nA} and μ_{nA} are the chemical potentials of neutrons and protons within heavy nuclei. These quantities are related by the Saha equation:

$$2\mu_n + 2\mu_p = \mu_\alpha, \tag{5.172}$$

$$2\mu_{nA} + 2\mu_{pA} = \mu_\alpha, \tag{5.173}$$

$$\mu_{nA} - \mu_{pA} = \mu_n - \mu_p = \hat{\mu}. \tag{5.174}$$

5.6.7 Baryon matter above nuclear density

We use the same supranuclear equation of state as that used by McAbee and Wilson [25]. In that work, heavy ion collisions of ^{57}La on ^{57}La were studied as a means to constrain the supernova equation of state. The electron fraction for ^{57}La ($Y_e = 0.41$) overlaps that of supernovae which ranges from $Y_e = 0.05$ to 0.50. It is important that laboratory experiments have been employed to constrain the equation of state.

As discussed in Chapter 2, this equation of state allows for the presence of pions. In thermal equilibrium the pion number densities are given by the Bose statistics:

$$N_i = \int_0^\infty \frac{d^3p}{h^3} \frac{1}{\exp\left[(\epsilon_\pi - \mu_i)/kT\right] - 1},$$
(5.175)

where ϵ_π is given by Eq. (5.140). Since ϵ_π depends on N_i through a dispersion relation, N_i occurs on both sides of Eq. (5.175).

If the pions are in chemical equilibrium with the other nuclear matter the reactions

$$p \rightarrow n + \pi^+, \qquad n \rightarrow p + \pi^-,$$
(5.176)

lead to the following constraints on the pion chemical potentials,

$$\mu_p = \mu_n + \mu_{\pi^\cdot}, \qquad \mu_n = \mu_p + \mu_{\pi^-}.$$
(5.177)

This implies

$$\mu_{\pi^\cdot} = -\mu_{\pi^-} = \mu_n - \mu_p.$$
(5.178)

The chemical potential of π^0 is zero.

In order to find μ_{π^-} and $\mu_{\pi^\cdot}$ one must have expressions for the nucleonic chemical potentials. The free energy per nuclear particle is taken to be separable into functions of ρ, Y_p, and T of the form

$$F = F_1(\rho) + F_2(\rho, Y_p) + F_3(\rho, T),$$
(5.179)

where the zero temperature, $Y_p = 0.5$ component is

$$F_1 = 8.79 \text{ MeV} + E_0 + \frac{1}{9}K_0\left[\frac{\eta^\Gamma - 1 - \Gamma(\eta - 1)}{\eta\Gamma(\Gamma - 1)}\right],$$
(5.180)

and the zero temperature asymmetry contribution is

$$F_2(\rho, Y_p) = \eta[16 \text{ MeV} + 72 \text{ MeV} (1 + 4\eta)^{-1}](1 - 2Y_p)^2.$$
(5.181)

Here, $\eta \equiv \rho/\rho_N$ where $\rho_N = 2.667 \times 10^{14}$ g cm^{-3} is the density of nuclear matter. The compressibility parameter is taken as $K_0 = 200$ MeV,

$\Gamma = 2.75$, and the binding energy per nucleon at nuclear density is $E_0 = -16$ MeV. These expressions for the zero temperature limit of the free energy F_1 and F_2 are taken from Muther *et al.* [31]. The extra 8.79 MeV added to F_1 corrects for the fact that our zero is with respect to iron nuclei, while theirs is with respect to free baryons.

The thermal contribution $F_3(\rho, T)$ is constructed as follows. One begins with a degenerate gas of quasi-nucleons (neutrons and protons) and delta particles. Their relative numbers are fixed by the equality of the chemical potentials. Only the thermal parts of the pressure and energy per baryon need be considered. Thus we write

$$F_3(\rho, T) = \Theta(\rho, t) - \Theta(\rho, 0), \qquad (5.182)$$

where

$$\Theta(\rho, t) = \sum_{i=N,\Delta} \int \frac{4\pi g_i dp_i p_i^2}{h^3} \left(\frac{\mu_i}{D_i} - kT \ln(D_i) \right), \qquad (5.183)$$

where g_i is the usual spin/isospin degeneracy factor and

$$D_i = \exp\left[(\epsilon_i - \mu_i)/kT\right] + 1 \qquad (5.184)$$

Here, the relativistic energies are

$$\epsilon_i = \sqrt{p_i^2 + (m_i^*)^2}, \qquad (5.185)$$

where the effective mass is

$$m_i^* = \frac{m_i}{1 + 0.27\rho/\rho_N}, \qquad (5.186)$$

with m_i the average rest mass of the nucleon or delta particle.

The chemical potentials μ_i are themselves also functions of ρ and T. They are determined by the condition of baryon number conservation

$$\rho N_A = \sum_{N,\Delta} \int \frac{4\pi g_i dp_i p_i^2}{h^3} \frac{1}{D_i}, \qquad (5.187)$$

where N_A is Avogadro's number and $\mu_N = \mu_\Delta$ is assumed.

The chemical potentials for neutrons and protons are found from the total free energy:

$$\mu_n = \frac{\partial F}{\partial Y_n}, \qquad \mu_p = \frac{\partial F}{\partial Y_p}. \qquad (5.188)$$

For the supranuclear equation of state, $Y_p = 1 - Y_n$ which implies

$$\mu_n = -\mu_p = \frac{4F_2(\rho, Y_p)}{(1 - 2Y_p)}. \qquad (5.189)$$

Thus, the nuclear symmetry energy per particle at nuclear density is 30.4 MeV.

The above constraints determine the pion chemical potentials. One also has the condition of charge neutrality which requires

$$Y_p + Y_{\pi^\cdot} - Y_{\pi^-} = Y_e, \tag{5.190}$$

where $Y_{\pi^\pm}$ is the net number of pions per baryon. An iteration method is used to solve simultaneously for $N_{\pi^\cdot}$, $\mu_{\pi^\cdot}$, Y_p, and $\hat{\mu}$.

The pion contribution to the energy density E_π and the pressure P_π is found from the usual Bose–Einstein statistics

$$E_\pi = \int \frac{d^3p}{h^3} \frac{\epsilon_\pi}{\exp\left[(\epsilon_\pi - \mu_i)/kT\right] - 1}, \tag{5.191}$$

and

$$P_\pi = \int \frac{d^3p}{h^3} \frac{(p_\pi/3)\partial\epsilon_\pi/\partial p_\pi}{\exp\left[(\epsilon_\pi - \mu_i)/kT\right] - 1}, \tag{5.192}$$

where ϵ_π is given by Eq. (5.140).

5.6.8 Numerical implementation of the equation of state

In numerical simulations, the equation of state is used to update temperature and pressure as the matter rest mass density ρ, matter internal energy ϵ_M, and electron fraction Y_e are evolved. To achieve this, a Newton–Raphson iteration technique is utilized to find the temperature consistent with a given ρ, ϵ_M, and Y_e. The temperature is updated in all zones at least once per iteration cycle. It is updated more often if the internal energy ϵ_M changes by more than some specified amount during a call to one of the routines in the operator splitting.

An estimate of the temperature and pressure is obtained between calls to the equation of state routine by means of a "specific heat" (C_V) and an "ideal index" Γ_{ideal} such that $\epsilon_M = C_V T$ and $P_M = (\Gamma_{ideal} - 1)\rho\epsilon_M$. During subroutine calls where ϵ_M changes in value, a new estimate of the temperature is thus easily obtained by

$$T \approx \frac{\epsilon_M}{C_V}. \tag{5.193}$$

Similarly, if the density changes from ρ_1 to ρ_2, then the new pressure can be approximated by

$$P_2 \approx P_1(\rho_2/\rho_1)^{\Gamma_{ideal}}. \tag{5.194}$$

5.7 Convection

Many papers have been written in recent years on the crucial roles played by convection to the success of the neutrino heating mechanism in a supernova explosion. Hence, it is important to have a convection algorithm which is both accurate and computationally efficient. Unfortunately, a precise three-dimensional theory for convection is difficult to implement. Below follows a description of the way in which various aspects of convection can be approximately accounted for in a one-dimensional spherical model.

There are two separate regions in which convection can be important. These are above and below the neutrinosphere radius R_ν. The neutrino sphere radius separates the star into two regions. Each becomes convectively unstable at different times. Below R_ν convection can be induced by a doubly diffusive phenomenon, similar to a *salt finger instability* [36]. The convective flows in this instability have been dubbed *neutron fingers* [37]. Outside R_ν quasi-Ledoux convection occurs [17, 30]. Details of the relativistic convection treatment are given below.

5.7.1 Mixing length theory

The basic elements of the mixing length theory employed in the Mayle and Wilson supernova model are as described in Cox and Giuli [13]. Depending upon the type of convection invoked, a convective velocity v_C and a mixing length λ_C can be determined. A convective diffusion coefficient can then be defined

$$D_C = f_C v_C \lambda_C, \tag{5.195}$$

where the dimensionless factor f_C takes into account the "angular distribution" of the convective motion and is of order unity.

The following phenomenological evolution equations are taken to represent the effects of convection:

$$\rho \frac{\partial Y_e}{\partial t} = \frac{\Gamma}{R^2} \frac{\partial}{\partial R} \left(a R^2 \rho D_C \frac{\partial Y_e}{\partial R} \right), \tag{5.196}$$

and

$$\rho \frac{\partial \epsilon_M}{\partial t} = \frac{\Gamma}{R^2} \frac{\partial}{\partial R} \left(a R^2 \rho D_C \zeta \frac{\partial \epsilon_M}{\partial R} \right), \tag{5.197}$$

where ρ is the rest mass matter density, ϵ_M is the matter internal energy, T is the material temperature, Y_e is the electron fraction, and ζ is a factor of order unity that corrects for the fact that in Eq. (5.197) we define the energy flux in terms of a gradient in temperature rather than in terms of an energy gradient. The effects of general relativity are taken into account by the factors Γ and a.

5.7.2 Convection phenomenology

The regions of the core that are unstable to convection can be identified by the following thought experiment. Imagine a blob of matter moving radially away from the center of the core. If it has become less dense than the ambient material at the new height, it will be driven further away from the center by buoyancy forces. A general criterion for determining an unstable region is that the following condition is satisfied:

$$\Delta \rho_B < \lambda_C \frac{d\rho}{dR}, \tag{5.198}$$

where $\Delta \rho_B$ is the change in the blob's density in moving outward by a distance λ_C, and $d\rho/dR$ is the radial density gradient in the core (in this section total derivatives refer to the stellar configuration). Different types of convection occur depending on the physical conditions of the blob's motion.

Quasi-Ledoux convection. The standard Ledoux convection (as is often employed in stellar evolution calculations) occurs when one considers a blob of fluid in pressure equilibrium without heat transfer or change in composition. For standard Ledoux convection the "composition" refers to the nuclear composition (e.g. ^{4}He, ^{12}C, ^{16}O, ^{24}Mg, etc.). In a supernova where nuclear reaction rates become large enough for the nuclear material to be in nuclear statistical equilibrium (NSE), the nuclear composition is uniquely determined by the density, temperature, and charge per baryon, Y_e. As a blob of material is transported in a supernova, its nuclear composition changes in response to the ambient conditions. That is, it is not constant.

Therefore, one can define a different type of convection called "quasi-Ledoux." Imagine now again that a blob moves in pressure equilibrium without heat transport. For quasi-Ledoux convection, the condition for a blob to become unstable (5.198) at the point $R + \lambda_C$ now becomes

$$\rho(P_B, S_B, Y_e^B) - \rho(P, S, Y_e) < 0, \tag{5.199}$$

where P is the ambient pressure, S is the entropy, and Y_e the ambient electron fraction. The subscript B denotes the same quantities for the blob. Expanding both terms in Eq. (5.199) about R gives

$$\frac{\partial \rho}{\partial S}\left(\Delta S_B - \lambda_C \frac{dS}{dR}\right) + \frac{\partial \rho}{\partial Y_e}\left(\Delta Y_e^B - \lambda_C \frac{dY_e}{dR}\right) + \frac{\partial \rho}{\partial S}\left(\Delta P_B - \lambda_C \frac{dP}{dR}\right) < 0. \tag{5.200}$$

The case of quasi-Ledoux convection assumes $\Delta S_B = 0$ and $\Delta P_B = \lambda_C dP/dR$, i.e. the blob moves with no heat exchange but in pressure equilibrium. In the region outside the neutrinosphere, the time scale for the blob electron fraction Y_e^B to equilibrate with the ambient material is on the order of a dynamic time scale (i.e. λ_C/v_C, the ratio of convective length scale to convective velocity). The change in Y_e^B can be parameterized by

$$\Delta Y_e^B = (1 - \alpha)\lambda_C \frac{dY_e}{dR}, \qquad (5.201)$$

where $0 \leq \alpha \leq 1$ is a dimensionless ratio of the dynamic to equilibrium neutrino interaction timescales. Using the above conditions, the condition of quasi-Ledoux convection becomes:

$$\left(\frac{\partial \rho}{\partial S}\right)_{P,Y_e} \lambda_C \frac{dS}{dR} + \alpha\left(\frac{\partial \rho}{\partial Y_e}\right)_{P,S} \lambda_C \frac{dY_e}{dR} > 0. \qquad (5.202)$$

Making use of the properties of the partial derivatives, this can be rewritten as

$$\left(\frac{\partial \rho}{\partial S}\right)_{P,Y_e} \left[\frac{dS}{dR} + \alpha\left(\frac{\partial \rho}{\partial Y_e}\right)_{P,S}\left(\frac{\partial S}{\partial \rho}\right)_{P,Y_e} \frac{dY_e}{dR}\right] > 0. \qquad (5.203)$$

Now,

$$\left(\frac{\partial \rho}{\partial Y_e}\right)_{P,S}\left(\frac{\partial S}{\partial \rho}\right)_{P,Y_e} = \left(\frac{\partial S}{\partial Y_e}\right)_{P,\rho}, \qquad (5.204)$$

and

$$\left(\frac{\partial \rho}{\partial S}\right)_{P,Y_e} = \frac{T}{C_P}\left(\frac{\partial \rho}{\partial T}\right)_{P,Y_e}, \qquad (5.205)$$

where C_P (> 0) is the specific heat capacity at constant pressure. Using Eqs. (5.204) and (5.205), Eq. (5.203) becomes

$$-\frac{T}{C_P}\left(\frac{\partial \rho}{\partial T}\right)_{P,Y_e}\left[\frac{dS}{dR} + \alpha\left(\frac{\partial S}{\partial Y_e}\right)_{P,\rho} \frac{dY_e}{dR}\right] < 0. \qquad (5.206)$$

To estimate the relative contributions of the terms dS/dR and dY_e/dR in Eq. (5.206) one can utilize the first law of thermodynamics applied to a free gas of electrons, positrons, free baryons, and photons,

$$dE = kTdS - Pd\left(\frac{m_N}{\rho}\right) + (\mu_e - \hat{\mu})dY_e, \qquad (5.207)$$

where E is the energy per baryon, m_N is the baryon mass, μ_e is the electron chemical potential, and $\hat{\mu}$ is the difference between the neutron and proton chemical potentials. Using Eq. (5.207) we have

$$\left(\frac{\partial S}{\partial Y_e}\right)_{P,\rho} = \frac{1}{kT}\left(\frac{\partial E}{\partial Y_e}\right)_{P,\rho} - \frac{(\mu_e - \hat{\mu})}{kT}. \qquad (5.208)$$

Both terms on the right-hand side of Eq. (5.208) are of order unity for the conditions within a supernova. Furthermore, the quantity Y_e only changes slightly from values just less than 1/2 to near 1/2 as R increases at early times after bounce. Hence, dY_e/dR is much smaller than dS/dR in the convective region. Thus, a neglect of the dY_e/dR term in Eq. (5.206) is justified. This is fortunate because one can estimate from Eq. (5.201) that the time scale for Y_e^B to come into equilibrium with the ambient material is within an order of magnitude of the time it takes for the blob to move the length of the convective region. Thus, it would be difficult to take into account properly the term involving dY_e/dR.

For reasonable equations of state the thermal density gradient is negative, i.e. $(\partial\rho/\partial T)_{P,Y_e} < 0$. Hence, from Eq. (5.206) we infer that the condition for quasi-Ledoux convection to occur is a negative entropy gradient, $dS/dR < 0$. This indeed happens behind the outgoing supernova shock. Convection in the region exterior to the neutrinosphere may develop in a manner that the fewest number of convective cells allowed by symmetry constraints form. Thus, in this model the mixing length for quasi-Ledoux convection is taken as the distance over which $dS/dR < 0$. That is, one takes

$$\lambda_C = R_{max} - R_{min},\qquad(5.209)$$

where R_{max} and R_{min} define the boundaries of the negative entropy gradient.

A convective velocity (v_C) is found from the relation between the convective kinetic energy and the work done by buoyant forces,

$$\frac{1}{2}\rho v_C^2 = \delta\rho g\lambda_C,\qquad(5.210)$$

where

$$g \equiv -\frac{1}{\rho}\frac{dP}{dR}, \qquad \delta\rho \equiv \lambda_C\left(\frac{\partial\rho}{\partial S}\right)_{P,Y_e}\frac{dS}{dR}.\qquad(5.211)$$

One can use Eqs. (5.210) and (5.211) to derive an expression for the convective velocity,

$$v_C = \lambda_C\left[2\frac{g}{\rho}\left(\frac{\partial\rho}{\partial S}\right)_{P,Y_e}\frac{dS}{dR}\right]^{1/2}.\qquad(5.212)$$

For the function f_C ($D_C = f_C v_v \lambda_C$) we use the ansatz,

$$f_C = 4\left(1 - \frac{R}{R_{max}}\right)\left(\frac{R}{R_{max}} - \frac{R_{min}}{R_{max}}\right)\left(1 - \frac{R_{min}}{R_{max}}\right)^{-1}.\qquad(5.213)$$

This phenomenology takes into account the fact that convection in this region may form with only a single convective cell moving in the radial direction (at both R_{min} and R_{max} the material is presumed to move perpendicular to the radius).

Neutron finger convection. In the delayed bounce mechanism for Type II supernovae the shock wave slows while trying to exit the iron core. It is then revived when neutrinos streaming from the proto neutron star heat the material behind the shock through nuclear absorption and inelastic scattering from electrons. Clearly, any process that increases the neutrino luminosity increases the likelihood that the slowed shock will start moving again and produce a supernova. Convection could bring significant amounts of proton rich material from within the inner core up to a region where electron capture is no longer blocked by neutrino degeneracy. This would increase the neutrino flux appreciably over what would otherwise be expected.

One type of convection which has proven crucial to the Mayle and Wilson's [44] supernova model is a doubly diffusive phenomenon. Following Smarr *et al.* [37], we will call this "neutron fingers." Doubly diffusive situations are characterized by having two spatial gradients, often temperature and composition, in the same direction as the gravitational field. One gradient is stabilizing and the other is destabilizing. In this way it is possible for a fluid to be unstable to convection even though the negative entropy gradient of the quasi-Ledoux criterion is not satisfied.

It is useful to consider ordinary salt fingers before proceeding to neutron fingers. Salt and temperature gradients are easier to visualize. Imagine a layer of hot, salty water residing over a layer of cold fresh water. Salty water is more dense than fresh so it should sink, but cold water is more dense than hot. Hence, one gradient is stabilizing while the other is destabilizing. Heat diffuses much more quickly than salt in water. The ratio of the thermal diffusivity in fresh water to that in salt water is ~100. If a fluid element of cold fresh water should rise up into the hot salty water it will quickly heat up. However, the hot fresh blob will be less dense that the hot salty water and it will continue to rise. The opposite is true, however, of a hot salty blob that enters cold fresh water. It becomes a cold salty blob that is more dense than the cold fresh water around it. Hence it commences to sink. The situation is clearly unstable. Ultimately, long thin fingers of hot, salty water penetrate the cold, fresh bottom layer and vice verse.

A similar initial arrangement exists in the proto neutron star. The collapse stops when the center reaches nuclear density. The homologous (velocity proportional to radius) inner core (of roughly 1/2 $M_\odot$) is falling at subsonic speeds. Hence, the information that the collapse has halted is transmitted as a relatively gentle sound wave. Outside the homologous core, however, the infall is supersonic. This causes a shock to form at the *sonic point*, i.e. the point at which the collapse velocity exceeds the speed of sound for the material.

The shock moves outward, raising the entropy of the outer mantle to three to four times that of the inner core. Neutrinos trapped in the inner core block electron capture and prevent significant deleptonization there. Y_e remains at about 0.3. The outer mantle is hotter and less dense, so neutrinos can escape more easily. This leaves room in the phase space for neutrinos released by electron capture, which drives Y_e down to 0.1. Thus, Y_e decreases with radius. The final picture is a sphere of about 1.5 $M_\odot$ with significant gradients in both entropy S and Y_e.

This situation is analogous to the salt finger instability. That is, high entropy material with a low Y_e lies on top of lower entropy material with a higher Y_e. The rapidly diffusing substance is again heat, transported in this case by neutrinos of all types ($\nu_e, \bar{\nu}_e, \nu_\mu, \bar{\nu}_\mu, \nu_\tau, \bar{\nu}_\tau$). The more slowly diffusing quantity is Y_e which depends upon only the ν_e and $\bar{\nu}_e$ neutrinos. In fact, as long as the neutrinos are not too degenerate, the diffusion of both species in the direction perpendicular to the radius tends to offset each individual species' effect on Y_e so that Y_e can only come into equilibrium slowly. This leads to the development of filaments of neutron rich material as depicted schematically on Figure 5.2. The importance of this phenomenon is that it allows the regions of neutrino-trapped material to convect up to near the surface. This increases the rate and efficiency of neutrino emission from the hot core. This is crucial to the revival of the slowed shock by neutrino heating.

One can model the neutron finger convection by assuming that a blob of material moves in thermal and pressure equilibrium, but not in chemical equilibrium with the surroundings. The condition for instability to occur

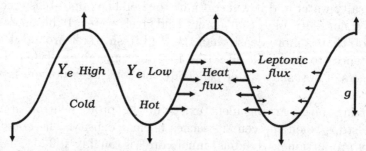

Fig. 5.2. Schematic illustration of the neutron finger convective instability.

[11] can then be written

$$\rho(P,T,Y_e^B) - \rho(P,T,Y_e) \approx \left(\frac{\partial \rho}{\partial Y_e}\right)_{P,T} \left(\Delta Y_e^B - \lambda_C \frac{dY_e}{dR}\right) < 0. \quad (5.214)$$

One can then take into account the change in Y_e^B as the blob moves in a similar manner to that for the quasi-Ledoux convection (5.201),

$$\Delta Y_e^B = (1 - \beta)\lambda_C \frac{dY_e}{dR}, \quad (5.215)$$

where $0 \leq \beta \leq 1$. Thus Eq. (5.214) can be rewritten

$$\beta\left(\frac{\partial \rho}{\partial Y_e}\right)_{P,T} \lambda_C \frac{dY_e}{dR} > 0. \quad (5.216)$$

In laboratory experiments of salt fingers, convective cells are observed [36] to develop as long thin structures parallel to the gravitational field. This suggests that for neutron fingers the length scale should be taken as the extent of the region over which the inequality (5.216) holds true. Hence, as in the case of quasi-Ledoux convection, one can take, $\lambda_C = R_{max} - R_{min}$. For reasonable equations of state $(\partial \rho / \partial Y_e)_{P,T} < 0$, so R_{max} and R_{min} can be identified as the region over which $dY_e/dR < 0$.

To estimate β in Eq. (5.216), one can estimate the ratio r of the rates of Y_e changing processes to those of energy transport by neutrino interactions in typical supernova conditions:

$$r = \frac{|N_{\nu e} - N_{\bar{\nu} e}|}{N_{\nu e} + N_{\bar{\nu} e} + N_{\nu \mu} + N_{\bar{\nu} \mu} + N_{\nu \tau} + N_{\bar{\nu} \tau}} < 0.1, \quad (5.217)$$

where N_i $(i = \nu_e, \bar{\nu}_e, \nu_\mu, \bar{\nu}_\mu, \nu_\tau, \bar{\nu}_\tau)$ are the number densities of various neutrino types. This means that changes is Y_e are small. Hence, one can take $\beta \approx 1$.

The convective velocity v_C in this case can be estimated by balancing the buoyancy force $(\delta \rho g)$ against the force due to neutrino viscosity,

$$F_{visc} = \eta_\nu \rho \nabla^2 \vec{V}, \quad (5.218)$$

where η_ν is the kinematic viscosity and $\vec{V}$ is the material three-velocity. Using the Eulerian momentum equation (cf. Eq. (2.22)) the convective velocity can be written

$$v_C = \frac{\delta \rho}{\rho} \frac{g}{\eta_\nu} (\lambda_F)^2, \quad (5.219)$$

with

$$\delta\rho = \lambda_C \left(\frac{\partial\rho}{\partial Y_e}\right)_{P,T} \frac{dY_e}{dR}, \tag{5.220}$$

and λ_F is a distance characteristic of the width of a neutron finger convective cell. This latter quantity can be estimated from dimensional analysis along with a linear stability analysis [19]

$$\lambda_F = \left(\frac{\kappa_\nu \eta_\nu}{g\xi d\ln t/dR}\right)^{1/4}, \tag{5.221}$$

where $\xi = -(\partial\ln\rho/\partial\ln T)_{P,Y_e}$ is the isobaric compressibility, κ_ν is the thermal diffusivity (i.e. thermal condictivity divided by the specific heat), and η_ν might be called the isobaric thermal expansion coefficient. Expressions for κ_ν and η_ν can be derived from kinetic theory. Simple estimates, adequate for this application are

$$\kappa_\nu = \frac{1}{3}\lambda_\nu c, \qquad \eta_\nu = \frac{1}{3}\lambda_\nu c\frac{\rho_\nu}{\rho}, \tag{5.222}$$

where λ_ν is the neutrino mean free path, ρ_ν is the neutrino mass energy density, ρ is the matter mass energy density, and c is the speed of light. From these expressions the convective velocity becomes

$$v_C = \frac{\delta\rho}{\rho}\left[\frac{g}{\xi}\frac{\rho}{\rho_\nu}\frac{T}{(dT/dR)}\right]^{1/2}. \tag{5.223}$$

Since neutron fingers are long thin structures, the multiplier f_C in the diffusion coefficient $(D_C = f_C v_C \lambda_C)$, which for the exterior quasi-Ledoux convection represents a geometrical factor (cf. Eq. (5.213)), becomes a neutron finger parameter which is used to make the calculated neutrino flux consistent with the observed flux from SN 1987A. From this, a value of $f_C = 0.02$ is deduced.

5.8 Model of a 20 M$_\odot$ supernova explosion

Here we summarize the main features of the specific calculation by Mayle and Wilson [22] of a 20 M$_\odot$ progenitor star as a model for supernova SN 1987A. The calculation began with the precollapse 20 M$_\odot$ model of Weaver and Woosley [41, 42] and Arnett *et al.* [1] which is specifically a model for SN 1987A. The core physics is independent of the blue or red nature of the supergiant, so the fact that SN 1987A was a blue giant is irrelevant to our discussion. The calculations followed the initial model through the core bounce, shock, late time heating, and collapse of the proto neutron star.

In Woosley *et al.* [47] the evolution of the bubble was followed in 90 exponentially increasing radial zones extending from the neutrinosphere to the shock. The mass resolution of zones at the neutrinosphere ($\sim$11 km) was $\sim 10^{-3}$ $M_\odot$. However, the density falls rapidly above the neutrinosphere such that the zonal mass resolution within the hot bubble at times and positions of interest was consistently 10^{-7} $M_\odot$ out to 200 km and 10^{-6} $M_\odot$ out to 1000 km. With this zoning, the flow of the neutrino heated material through the bubble was well resolved.

The evolution of the star was followed to 18 s after bounce. As the shock moved out through the star, the zoning became poorer. This led to some uncertainty in the calculations near the neutrinosphere where the temperature gradient becomes steep late in the calculation. Nevertheless, the zoning was adequate for a study of the nucleosynthesis above the proto neutron star.

In these calculations, the matter equation of state included contributions from electrons, positrons, photons, free baryons, helium, representative heavy nuclei, pions, and kaons as described herein. The distributions in energy, space, and time were followed for electron, anti-electron, μ, and τ neutrinos. A simple nine isotope network followed the energy released

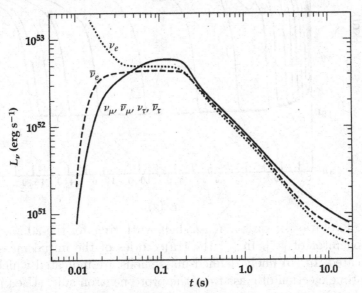

Fig. 5.3. Sketch of average neutrino luminosity versus time for the 20 $M_\odot$ supernova progenitor of [47]. (Used by permission of *The American Astronomical Society*.)

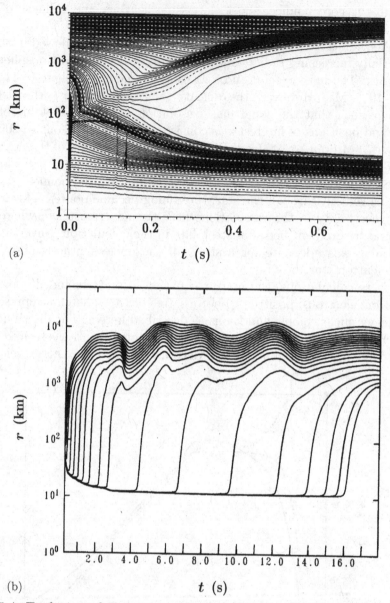

(a)

(b)

Fig. 5.4. Evolution of various mass shells with time for the 20 M$_\odot$ supernova progenitor of [47]. In (a) the trajectories of the initial mass shells
are indicated. In (b) finely defined mass shells are indicated which show
the late time ejection of mass from the proto neutron star. (Used by permission of *The American Astronomical Society*.)

by thermonuclear reactions. The hydrodynamics and neutrino flow were evaluated as described in this chapter with all of the important neutrino–matter interactions included (cf. [22]). Neutrino capture on free nucleons is the most important source of energy for the explosion. However, neutrino annihilation with anti-neutrinos also contributes 10–15%, and neutrino–electron scattering contributes 15–25% to the explosion energy. Capture and inelastic scattering on heavy nuclei also contribute to a lesser extent.

Convection was modeled in two separate regions as described in this chapter, beneath the neutrinosphere and above it. Below the neutri-nosphere a convective instability requires the determination of the neutrino distributions functions. This is a critical component of the supernova model. The mixing length convection model [22] used in the spherical code calculations has been shown to be adequate up to this time. Later, after the bubble becomes very hot, the convection picture becomes more complicated and mixing length theory is no longer adequate [8].

A sketch of the luminosity of emerging neutrinos from that calculation is given in Figure 5.3. The delayed neutrinos arriving after ≈0.1 s cause the rejuvenation of the shock and outward expansion as evidenced in the mass contours given in Figure 5.4.

An interesting feature of this result is that the high entropy bubble formed by the neutrino heating seems to provide an ideal site for the synthesis of heavy nuclei by rapid neutron capture (*r*-process). Nucleosynthesis occurs as photodissociated matter is ejected from the proto neutron star and passes through the high entropy bubble. Figure 5.5 summarizes

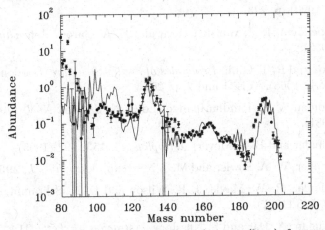

Fig. 5.5. Final integrated *r*-process abundances (line) from the 20 M⊙ supernova progenitor of [47] compared with the solar system *r*-process abundances. (Used by permission of *The American Astronomical Society*.)

the agreement between observed solar system abundances [18] and the calculated abundances for material which has passed through the bubble. Although these results look promising, elements with $A \sim 90$ were significantly overproduced by a factor of more than 100. Furthermore, neutrino–nucleus interaction processes have been shown [13, 26, 28, 34] to hinder the r-process by decreasing the neutron-to-seed abundance ratio (although they can help to smooth the final abundance pattern). Recent calculations with the improved neutrino annihilation representation [35] give much more production of high mass nuclei. This should help to overcome the difficulties discussed in [13].

References

[1] W. D. Arnett, J. N. Bahcall, R. P. Kirshner and S. E. Woosley, *Annu. Rev. Astron. Astrophys.*, **27**, 629 (1989).

[2] E. Baron and J. Cooperstein, *Astrophys. J.*, **353**, 597 (1990).

[3] H. A. Bethe, *Rev. Mod. Phys.*, **62**(4), 801 (1990).

[4] H. A. Bethe and J. R. Wilson, *Astrophys. J.*, **295**, 14 (1995).

[5] R. L. Bowers and J. R. Wilson, *Astrophys. J. Suppl. Ser.*, **50**, 115 (1982).

[6] R. L. Bowers and J. R. Wilson, *Numerical Modeling in Applied Physics and Astrophysics* (Jones and Bartlett, Boston, MA, 1991).

[7] L. S. Cheng and G. E. Cooper, *J. Comput. Phys.*, **6**, 1 (1970).

[8] S. A. Colgate, M. Herant and M. Benz, *Phys. Rep.*, **227**, 157 (1993).

[9] J. Cooperstein and E. Baron, in *Supernovae*, ed. A. Petschek (Springer, Berlin, 1990), p. 223.

[10] J. Cooperstein, L. J. van den Horn and E. A. Baron, *Astrophys. J.*, **309**, 653 (1986).

[11] J. P. Cox and R. T. Giuli, *Principles of Stellar Structure* (Gordon & Breach, New York, 1968), Vols. 1 and 2, p. 267.

[12] B. Friedman, V. R. Pandharipande and Q. N. Usmani, *Nucl. Phys. A*, **372**, 483 (1981).

[13] G. M. Fuller and B. S. Meyer, *Astrophys. J.*, **453**, 792 (1995).

[14] G. M. Fuller, W. A. Fowler and M. J. Newman, *Astrophys. J.*, **293**, 1 (1985).

[15] G. M. Fuller, R. W. Mayle, J. R. Wilson and D. N. Schramm, *Astrophys. J.*, **322**, 795 (1987).

[16] J. Goodman, A. Dar and S. Nussinov, *Astrophys. J. Lett.*, **314**, 7L (1987).

[17] M. Herant, M. Benz and S. A. Colgate, *Astrophys. J.*, **395**, 642 (1992).

[18] F. Käppeler, H. Beer and K. Wisshak, *Rep. Prog. Phys.*, **52**, 945 (1989).

[19] P. F. Linden and J. S. Turner, *J. Fluid Mech.*, **67**, 1 (1975).

[20] G. J. Mathews and R. A. Ward, Neutron capture processes in astrophysics, *Rep. Prog. Phys.*, **48**, 1371 (1985).

[21] M. May and R. H. White, *Methods Comput. Phys.*, **73**, 219 (1967).

[22] R. W. Mayle and J. R. Wilson, *Astrophys. J.*, **334**, 909 (1988).

[23] R. W. Mayle and J. R. Wilson, in *Supernovae*, ed. S. E. Woosley (Springer, New York, 1990), p. 333.

[24] R. W. Mayle, M. Tavani and J. R. Wilson, *Astrophys. J.*, **418**, 398 (1993).

[25] T. L. McAbee and J. R. Wilson, *Nucl. Phys. A*, **576**, 626 (1994).

[26] G. C. McLaughlin, G. M. Fuller and J. R. Wilson, *Astrophys. J.*, **472**, 440 (1996).

[27] G. E. B. Messer, A. Mezzacappa, S. W. Bruenn and M. W. Guidry, *Astrophys. J.*, **507**, 353 (1998).

[28] B. S. Meyer, *Astrophys. J. Lett.*, **449**, 55 (1995).

[29] B. S. Meyer, W. M. Howard, G. J. Mathews, S. E. Woosley and R. D. Hoffman, *Astrophys. J.*, **399**, 656 (1992).

[30] D. S. Miller, J. R. Wilson and R. W. Mayle, *Astrophys. J.*, **415**, 278 (1993).

[31] H. Muther, M. Prakash and T. L. Ainsworth, *Phys. Lett. B*, **199**, 469 (1988).

[32] K. Otsuki, H. Tagoshi, T. Kajino and S. Wanajo, *Astrophys. J.*, **533**, 4240 (2000).

[33] Y.-Z. Qian, W. C. Haxton and S. E. Woosley, *Astrophys. J.*, **471**, 331 (1996).

[34] Y.-Z. Qian, W. C. Haxton, K. Langanke and P. Vogel, *Phys. Rev.*, **C55**, 1532 (1997).

[35] J. A. Salmonson and J. R. Wilson, *Astrophys. J.*, **517**, 859 (1999).

[36] E. Schroeder, H. Strommel, D. Menzel and W. Sutcliff, *J. Geophys. Res.*, **64**, 363 (1959).

[37] L. Smarr, J. R. Wilson, R. T. Barton and R. L. Bowers, *Astrophys. J.*, **246**, 515 (1981).

[38] K. Sumiyoshi, M. Terasawa, G. J. Mathews and T. Kagino, *Astrophys. J.*, **562**, 880 (2001).

[39] D. L. Tubbs, *Astrophys. J.*, **231**, 846 (1979).

[40] D. L. Tubbs and D. N. Schramm, *Astrophys. J.*, **201**, 467 (1975).

[41] T. A. Weaver and S. E. Woosley, *Phys. Rep.*, **227**, 65 (1993).

[42] T. A. Weaver, S. E. Woosley and N. Langer, *Astrophys. J.*, **411**, 823 (1994).

[43] J. R. Wilson, *Astrophys. J.*, **163**, 209 (1971).

[44] J. R. Wilson and R. W. Mayle, *Phys. Rep.*, **227**, 97 (1993).

[45] J. R. Wilson, R. Couch, S. Cochran, Z. J. Leblanc and Z. Barkat, *Ann. NY Acad. Sci.*, **262**, 54 (1975).

[46] S. E. Woosley, private communication (1986).

[47] S. E. Woosley, J. R. Wilson, G. J. Mathews, R. D. Hoffman and B. S. Meyer, *Astrophys. J.*, **433**, 229 (1994).

[48] S. Yamada, H.-T. Janka and S. Hideyuki, *Astron. Astrophys.*, **344**, 533 (1999).

6

Axially symmetric relativistic hydrodynamics

6.1 Systems with a fixed metric

For many problems with axial symmetry (e.g. accretion) the metric is dominated by an inner compact neutron star or black hole. The matter flow of interest, however, contributes negligibly to the metric. Hence, the metric can be treated as a fixed background. This eliminates the need to solve for the dynamics of the gravitational field and greatly simplifies the problem. We consider such problems first.

In a sense, hydrodynamics with respect to a fixed background is equivalent to doing special relativity in a fixed geometry. The metric itself could be of Kerr, Schwarzschild, conformally flat, or even post-Newtonian form. It matters not whence the metric arose nor whether it is an exact solution to the Einstein field equations.

6.1.1 Kerr metric

In cylindrical t, R, Z, ϕ coordinates the Kerr metric can be written

$$
g^{\mu\nu} = \begin{pmatrix} -\left[1 + \frac{(2mr(a^2+r^2)}{\rho^2\Delta}\right] & 0 & 0 & \frac{2amr}{\rho^2\Delta} \\ 0 & \frac{\Delta}{\rho^2}\left(\frac{Z^2}{\Delta} + \frac{R^2}{r^2}\right) & \frac{Rz}{\rho^2}\left(\frac{\Delta}{r^2} - 1\right) & 0 \\ 0 & \frac{Rz}{\rho^2}\left(\frac{\Delta}{r^2} - 1\right) & \frac{\Delta}{\rho^2}\left(\frac{r^2}{\Delta} - \frac{z^2}{r^2}\right) & 0 \\ \frac{2amr}{\rho^2\Delta} & 0 & 0 & \frac{1}{\rho^2}\left(\frac{r^2}{R^2} - \frac{a^2}{\Delta}\right) \end{pmatrix},
$$

(6.1)

165

where a is the specific angular momentum $a \equiv J/M^2$, and the other parameters are defined as follows: $r^2 \equiv R^2 + Z^2$; $\Delta \equiv r^2 - 2Mr + a^2$; and $\rho^2 \equiv r^2 + a^2 Z^2/r^2$.

The volume factor is written slightly differently here,

$$g \equiv \sqrt{-\det(g_{\mu\nu})} = R\left(1 + \frac{a^2 Z^2}{r^4}\right). \tag{6.2}$$

The axisymmetric hydrodynamic equations of motion [3] are then,

$$\dot{D} + \frac{1}{g}\frac{\partial}{\partial R}\left(DV^R g\right) + \frac{\partial}{\partial Z}\left(DV^Z g\right) = 0, \tag{6.3}$$

$$\dot{E} + \frac{1}{g}\frac{\partial}{\partial R}\left(EV^R g\right) + \frac{\partial}{\partial Z}\left(EV^Z g\right)$$
$$+ P\left\{\frac{\partial U^t}{\partial t} + \frac{1}{g}\left[\frac{\partial}{\partial R}\left(U^R g\right) + \frac{\partial}{\partial Z}\left(U^Z g\right)\right]\right\} = 0, \tag{6.4}$$

$$\dot{S}_i + \frac{1}{g}\frac{\partial}{\partial R}\left(S_i V^R g\right) + \frac{\partial}{\partial Z}\left(S_i V^Z g\right) + \frac{\partial P}{\partial x^i} + \frac{1}{2}\frac{\partial g^{\alpha\beta}}{\partial x^i}\frac{S_\alpha S_\beta}{S^t} = 0. \tag{6.5}$$

6.1.2 Accretion shocks

As an illustration, we describe the formation of shocks during accretion onto a Kerr black hole. Accretion onto a black hole was studied in [6]. The problem was to find which parameters for the black hole and the accreting gas led to the formation of a shock wave external to the black hole. The rotation was parametrized by the hole's specific angular momentum a, and the matter flow was parameterized by the specific angular momentum of material at the grid boundary, $A \equiv U_\phi(r_{max}/M)$. At the waist of an extreme ($a = 1$) Kerr black hole, the lowest gas velocity that can fall directly in is $U_\phi = 2.0$. Table 6.1 shows the results for shock formation.

Illustrations of the flow velocities for selected calculations are shown in Figure 6.1 from [6]. In most cases the gas falls into the hole with no appreciable heating. However, when material is corotating with the metric, shock waves are formed if the specific angular momentum of the black hole is close to its maximum value and if the specific angular momentum of the infalling material U_ϕ/M is greater than 2. For counter-rotation, $U_\phi < 0$, the matter is pulled into the black hole unless the counter-rotation velocity is extremely large, $U_\phi \ll 0$.

Table 6.1   Summary of accretion flow into a Kerr black hole

a	U_ϕ	Γ	Flow characteristics
0	0	3/2	smooth accretion
0	0	4/3	smooth accretion
0	0	4/3	smooth accretion
0.714	0	3/2	stationary shocks
1	3	3/2	smooth accretion
1	2	4/3	smooth accretion
1	3	4/3	shocks plus vortex behind shocks
1	4	4/3	shocks plus vortex behind shocks
1	−2	4/3	smooth accretion
1	−4	4/3	smooth accretion

6.1.3 Kerr accretion with magnetized gas

The Kerr accretion calculations described in the previous section have been extended [8–10] to a study of the accretion of magnetized gas. To see how this was done, let us first address how relativistic magnetohydrodynamics can be carried out in axially symmetric systems.

For axially symmetric systems, the problem simplifies if the magnetic field is represented [5] by H_ϕ and A_ϕ, where H_ϕ is the magnetic field in the ϕ direction, and A_ϕ is the ϕ component of the vector potential.

Magnetic fields are easily entered in relativity by adding the electromagnetic stress energy tensor,

$$T_{\mu\nu} = T_{\mu\nu}^{Fluid} + T_{\mu\nu}^{EM}, \tag{6.6}$$

where

$$T_{\mu\nu}^{EM} = \frac{1}{4\pi}(g_{\alpha\mu}F^{\mu\lambda}F_{\alpha\beta} + \frac{g_{\alpha\beta}}{4}F^{\mu\lambda}F_{\mu\lambda}), \tag{6.7}$$

and as usual, the electromagnetic tensor $F_{\mu\nu}$ can be related to a vector potential A_ν,

$$F_{\mu\nu} = \frac{\partial A_\nu}{\partial x^\mu} - \frac{\partial A_\mu}{\partial x^\nu}. \tag{6.8}$$

The nonvanishing spatial components of $F_{\mu\nu}$ are thus,

$$F_{rz} = H_\phi, \qquad F_{r\phi} = \frac{\partial A_\phi}{\partial r}, \qquad F_{z\phi} = \frac{\partial A_\phi}{\partial z}. \tag{6.9}$$

Then, from the assumption of perfect conductivity, $U^\mu F_{\mu\nu} = 0$, the spacetime components can be obtained:

$$F_{tr} = V^z H_\phi + V^\phi F_{r\phi}, \tag{6.10}$$

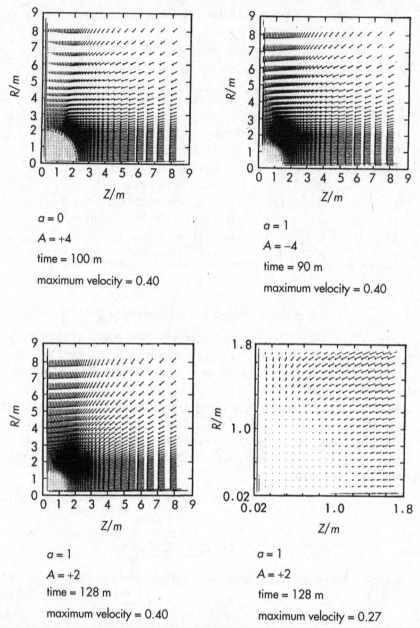

Fig. 6.1. R–Z projection of the three-velocity vector V^i for material flowing into a Kerr black hole, from [6]. Plots are labeled with various values of the black hole specific angular momentum $a = 0$ or 1 (extreme Kerr), and the accreting material specific angular momentum A defined as U_ϕ at the boundary radius divided by the mass of the black hole. The length of the arrows is scaled so that the maximum vector length is that indicated below each plot. (Used by permission of *The American Astronomical Society*.)

$$F_{tz} = V^\phi F_{z\phi} - V^r H_\phi, \tag{6.11}$$

$$F_{t\phi} = -V^r F_{r\phi} - V^z F_{z\phi} = \frac{\partial A_\phi}{\partial t}. \tag{6.12}$$

The time evolution of H_ϕ then arises from Maxwell's equation

$$F_{rz;t} + F_{tr;z} + F_{zt;r} = 0, \tag{6.13}$$

which gives

$$\frac{\partial H_\phi}{\partial t} = \frac{\partial}{\partial z}\left(V^\phi \frac{\partial A^\phi}{\partial r}\right) - V^z H_\phi - \frac{\partial}{\partial r}\left(V^\phi \frac{\partial A^\phi}{\partial z}\right) + V^r H_\phi. \tag{6.14}$$

The momentum equation must also be modified for the effects of a magnetic field on the matter (e.g. magnetic braking). These effects enter through an an additional term $J^\mu F_{\mu j}$, where the current J^μ is written,

$$J^\mu = \nabla_\nu F^{\mu\nu} = \frac{1}{g}\frac{\partial}{\partial x^\nu}\left(g F^{\mu\nu}\right). \tag{6.15}$$

The momentum equation (6.5) is then modified to become,

$$\dot{S}_j + \cdots + J^\mu F_{\mu j} = 0. \tag{6.16}$$

For low density magnetized gas it is important to include properly the inertia of the magnetic field in the momentum equation. To do this we identify the terms in $J^\mu F_{\mu j}$ that multiply the velocity acceleration $\dot{U}_j$. We call these coefficients $F(j)$. Then, we can rewrite the magnetohydrodynamic momentum equation (6.16) as

$$\dot{S}_j + \cdots + \dot{U}_j F(j) + (J^\mu F_{\mu j} - \dot{U}_j F(j)) = 0. \tag{6.17}$$

When doing the operator splitting for the numerical evaluation, one can then collect these terms together and solve for $\dot{U}_j$,

$$\dot{U}_j + \frac{(J^\mu F_{\mu j} - \dot{U}_j F(j))}{D + \Gamma E + F(j)} = 0, \tag{6.18}$$

so that the operator split $\dot{S}_j$ term becomes

$$\dot{S}_j + \frac{(J^\mu F_{\mu j} - \dot{U}_j F(j))}{1 + F(j)/(D + \Gamma E)} = 0. \tag{6.19}$$

For example, the relevant operator split radial momentum equation in spherical coordinates becomes

$$\dot{S}_j + \frac{1}{Q}(J^\mu F_{\mu j} - \dot{U}_j F(j)) = 0, \tag{6.20}$$

where

$$Q = 1 + \frac{1}{(D + \Gamma E)W}\left[g^{rr}g^{\theta\theta}H_\phi^2 + \frac{(g^{tt}g^{\phi\phi} - g^{\phi t}g^{\phi t})}{g^{tt}}\left(\frac{\partial A_\phi}{\partial r}\right)^2\right]. \quad (6.21)$$

This procedure normalizes the electromagnetic force by the direction dependent ratio of the inertial magnetic energy density to the inertial mass energy density. This keeps the system well behaved for magnetic fields with energy densities comparable to or greater than the matter mass energy density (i.e. near equipartition).

6.1.4 Magnetohydrodynamics results around a Kerr black hole

The accretion onto a Kerr black hole can be parameterized by a, U_ϕ, and H, where H is the ratio of magnetic field energy density to the gas density at the outside of the calculational grid. The field is greatly amplified if a and/or $U_\phi \neq 0$. In this case, strong shear develops in the flow, leading to large amplification of the magnetic field. Even with $a = 1$, $U_\phi = 1$, and $H = 0.0001$, complex flow results [2, 8, 9, 10] as shown in Figure 6.2.

One point of interest is the electric charge which accumulates in the black hole. The charge is determined in the calculation by evaluating the radial electric field at the surface of the black hole (Gauss' law). Figure 6.3 shows how the black hole charge evolves with time along with the fields near the equator of the black hole. The charge rises to a fixed value as the flow settles into a steady state. The conducting gas contains a total charge nearly equal to, but of opposite sign to, the black hole's charge (see Figure 6.3).

6.2 Rotating stars

We now consider axisymmetric systems in which the metric cannot be treated as a simple Kerr or Schwarzschild background, but must be solved along with the matter equations. The simplest example of this is the interior equilibrium configuration of axially symmetric rotating stars.

6.2.1 Rotating stars

Spherical supermassive stars are unstable because the effective pressure index in the interior approaches $\Gamma = 4/3$. This pressure response is too weak to resist the increased gravitational attraction that general relativity induces. A study [10] was made of rotating stars using the metric introduced by Bardeen and Wagoner [1],

$$ds^2 = B^2 e^{2F}[e^{2C}(dR^2 + dZ^2) + R^2(d\phi - \omega dt)^2)] - e^{-2F}dt^2. \quad (6.22)$$

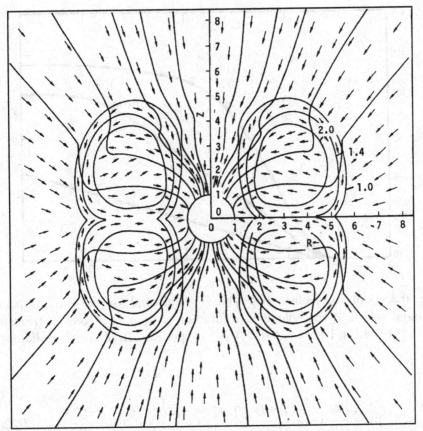

Fig. 6.2. Computed poloidal field lines [9] for the in-fall of material with $A = 1$ onto an extreme $a = 1$ Kerr black hole with $H = 0.0001$. Closed looping lines are contours of $A = 1, 1, 4, 2$. Arrows indicate the direction of material flow. In the region of high A the magnitudes of the velocity are 2–5 times less than the length would indicate and the velocity goes to zero near the hole. (Used by permission of *North Holland*.)

The Einstein field equations lead to the following conditions on the metric variables:

$$\nabla \cdot (B\nabla F) = -4\pi B^3 e^{2C+2F}[\rho(1+\epsilon)(1+v^2) + P(3-v^2)]/(1-v^2)$$
$$- R^2 B^3 \nabla\omega \cdot \nabla\omega e^{4F}/2, \tag{6.23}$$

$$\nabla \cdot (R^2 B^3 e^{4F}\nabla\omega) = -8\pi R B^4 v[\rho(1+\epsilon) + P]e^{4F+2C}/(1-v^2), \tag{6.24}$$

$$\nabla \cdot (R\nabla B) = 16\pi P R B^3 e^{2C+2F}, \tag{6.25}$$

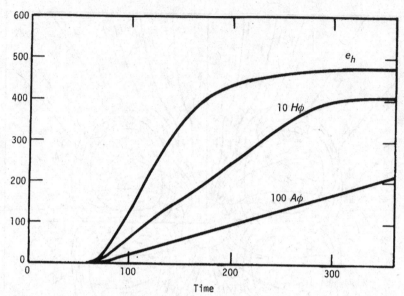

Fig. 6.3. Computed black hole charge, equitorial vector potential, and toroidal field near the hole [9] as a function of time for a Kerr metric with $a = 1$, $A = 1$, and $H = 0.0001$. (Used by permission of *North Holland*.)

and

$$\frac{\partial C}{\partial R} = \left[\left(1 + \frac{R}{B}\frac{\partial B}{\partial R}\right)S_1 + \frac{R}{B}\frac{\partial B}{\partial Z}S_2\right] \bigg/ \left[\left(1 + \frac{R}{B}\frac{\partial B}{\partial R}\right)^2 + \left(\frac{R}{B}\frac{\partial B}{\partial Z}\right)^2\right], \quad (6.26)$$

where

$$S_1 \equiv \frac{R}{4}\left\{\left(\frac{\partial \log B}{\partial Z}\right)^2 - \left(\frac{\partial \log B}{\partial R}\right)^2 + \frac{1}{2B}\frac{\partial^2 B}{\partial R^2} - \frac{1}{2B}\frac{\partial^2 B}{\partial Z^2} + \left(\frac{\partial F}{\partial R}\right)^2 \right.$$
$$\left. - \left(\frac{\partial F}{\partial Z}\right)^2 - R^2 B^2 e^{4F}\left[\left(\frac{\partial \omega}{\partial R}\right)^2 - \left(\frac{\partial \omega}{\partial Z}\right)^2\right]\right\}, \quad (6.27)$$

and

$$S_2 \equiv R\left(\frac{1}{B}\frac{\partial^2 B}{\partial R \partial Z} - \frac{2}{B^2}\frac{\partial B}{\partial R}\frac{\partial B}{\partial Z} + 2\frac{\partial F}{\partial R}\frac{\partial F}{\partial Z} - \frac{R^2 B^2 e^{4F}}{2}\frac{\partial \omega}{\partial R}\frac{\partial \omega}{\partial Z}\right). \quad (6.28)$$

The equation for hydrostatic equilibrium can be written

$$\frac{\partial P}{\partial x^i} + \frac{1}{2}\frac{\partial g^{\alpha\beta}}{\partial x^i}U_\alpha U_\beta[\rho(1 + \epsilon) + P] = 0. \quad (6.29)$$

The solution for a rotating star can be found by choosing [7, 10] a parameterized density distribution given by

$$\rho = \rho_0(a + aY + bY^2)e^{-Y}, \tag{6.30}$$

where

$$Y \equiv [R^n + (\alpha Z)^n]^{1/2}. \tag{6.31}$$

Starting with initial guesses for the fields B, F, ω, and C, Eqs. (6.23)–(6.26) are solved for a next approximation. Then Eq. (6.29) is integrated to find values of U_ϕ and ϵ consistent with the fields. The field quantities are then resolved with the new U_ϕ and ϵ. The process is repeated until convergence is achieved.

A parameter γ can be defined which gives a measure of the strength of the gravitational field,

$$\gamma = 1 - e^{-F_c}, \tag{6.32}$$

where F_c is the central value of F. (Note that γ is not related to the $\sqrt{\det(\gamma_{ij})}$ as in previous chapters.) The binding energy versus γ is shown in Figure 6.4 for several values of the shape parameter α. Unless differential rotation enhances stability, only stars with α 2 are stable. At $\alpha = 2$ the binding energy is 1% of the mass and so these stars might be stable enough to be of interest. In Figure 6.5 we show the surfaces of $g_{tt} = 0$. At this surface the frame dragging velocity becomes equal to light speed. This defines regions in which the matter cannot sit still.

In Figure 6.6 the binding energy for a $\Gamma = 2$ star (which crudely represents a neutron star) is presented for various values of α and γ. For low γ the shape is not very important. For $\alpha = 2$, which perhaps gives the greatest stable rotation, the maximum binding energy is increased by 40%.

6.2.2 Magnetic rotating stars

These rotating star calculations have been extended [10] to model magnetized stars. Magnetic fields were introduced by the method described in the section on Kerr accretion, i.e by representation of the field by the H_ϕ magnetic field component and the vector potential component A_ϕ. The comoving electric field is set to zero as before by the condition

$$U^\mu F_{\mu\nu} = 0. \tag{6.33}$$

The electromagnetic energy tensor is evaluated and then the field gravitational equations given in the previous section are augmented by

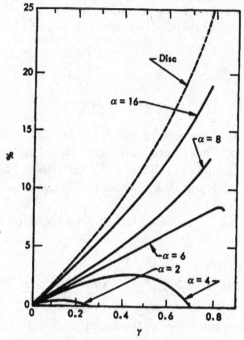

Fig. 6.4. Binding energy in percent of rest mass versus the gravity parameter γ for several shape factors α from [7]. (Used by permission of *The American Astronomical Society.*)

the appropriate magnetic terms. The equation for hydrostatic equilibrium (6.29) is modified by the magnetic field to be

$$\frac{\partial P}{\partial R} + \frac{1}{2}\frac{\partial g^{\alpha\beta}}{\partial R}U_\alpha U_\beta[\rho(1+\epsilon) + P] + J^\alpha F_{\alpha R} = 0, \qquad (6.34)$$

$$\frac{\partial P}{\partial Z} + \frac{1}{2}\frac{\partial g^{\alpha\beta}}{\partial Z}U_\alpha U_\beta[\rho(1+\epsilon) + P] + J^\alpha F_{\alpha Z} = 0. \qquad (6.35)$$

The method of solution is the same as for the nonmagnetized stars except that only two equations for equilibrium are available so that either the internal energy ϵ is set to zero and U_ϕ and H_ϕ are solved for, or one can also select a magnetic field configuration and then solve for ϵ and U_ϕ. For a cold neutron star, ϵ would be a function of ρ and so one could solve for U_ϕ and H_ϕ. To study equilibrium star configurations the initial density profiles of Eq. (6.30) were used.

In Figure 6.7 the results [10] of an $\epsilon = 0$, cold, low pressure star calculation are shown. For stars which are completely magnetically supported, the binding energies are substantial even for spherical stars.

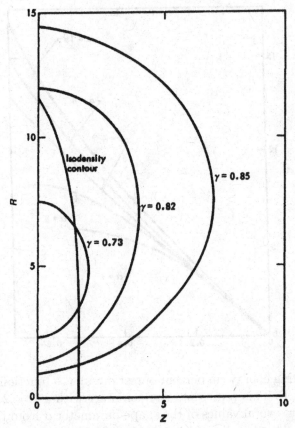

Fig. 6.5. Contours of $g_{tt} = 0$ for stars with a shape factor of $\alpha = 6$ for several values of the gravity parameter γ from [7]. (Used by permission of *The American Astronomical Society.*)

In order to make dynamic calculations of magnetized stars that initially have insufficient rotation and/or magnetic field, the equilibrium equations (6.34) and (6.35) are augmented by the dynamic hydrodynamic terms discussed earlier in this chapter. A density configuration is chosen at the start of the calculation and the stars are assigned ϵ values so as to be isentropic. A poloidal field is introduced. The field equations are solved to advance in time, and at each hydro cycle new gravitational fields are generated. Gravitational radiation is ignored.

During a run, the emitted gravitational radiation is estimated by

$$E = \frac{1}{M} \int (\dddot{Q})^2 dt, \qquad (6.36)$$

and E is found to be extremely small, $< 10^{-6}$.

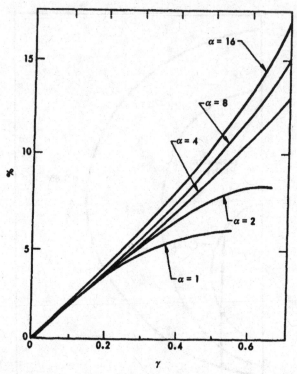

Fig. 6.6. Binding energy (in percent of rest mass) as a function of equation of state index Γ. This represents neutron stars with a $\Gamma = 2$ equation of state index for several values of the shape parameter α from [7]. (Used by permission of *The American Astronomical Society*.)

During the collapse of a magnetized rotating star, differential rotation develops which leads to greatly enhanced magnetic field energy. The generation of high magnetic field energy leads to the formation of a jet along the rotation axis as illustrated in Figure 6.7 from [10].

6.3 Systems with a dynamic metric

In systems that have a symmetry, such as axial symmetry, the full Einstein equations become simple enough that the full equations are readily solvable. An example of an axially symmetric system with no rotation is the head-on collision of two neutron stars [11, 12]. One reason for interest in this system is that it is a good test bed in which to scrutinize methods for computing binary neutron star systems. In particular, in this section we will examine this system as a means to explore the validity of the conformally flat approximation introduced in the next chapter.

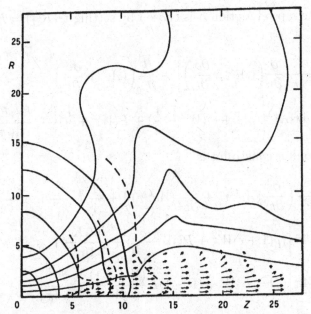

Fig. 6.7. Stellar configuration of a magnetized star just after an axial jet has formed, from [10]. The arrows indicate velocities in the (R, Z) plane. The longest vector represents a velocity of 0.25. The closely nested curves are isodensity curves. The dashed lines are isotoroidal magnetic field contours. The poloidal magnetic field lines start on the equator and go up and out. (Used by permission of *North Holland*.)

For the axisymmetric head-on collision problem, the ADM metric can be chosen as,

$$ds^2 = -\alpha dt^2 + A^2 \left[(\beta_r dt + dr)^2 + (\beta_z dt + dz)^2 \right] + A^2 e^{2C} r^2 d\phi^2. \quad (6.37)$$

To put the metric in this simplified form, the shift vector β_r, β_z is chosen so that $g_{rz} = 0$ and $\dot{g}_{rr} = \dot{g}_{zz}$. This requirement leads to the constraint equations,

$$\frac{\partial^2 \beta^z}{\partial r^2} + \frac{\partial^2 \beta^r}{\partial z^2} = \frac{\partial}{\partial r} \left[\alpha(K^r{}_r - K^z{}_z) \right] + 2\frac{\partial}{\partial z} \left(\alpha K^r{}_z \right), \quad (6.38)$$

and

$$\frac{\partial^2 \beta^z}{\partial r^2} + \frac{\partial^2 \beta^r}{\partial z^2} = -\frac{\partial}{\partial z} \left[\alpha(K^r{}_r - K^z{}_z) \right] + 2\frac{\partial}{\partial r} \left(\alpha K^r{}_z \right). \quad (6.39)$$

As usual, the lapse function α is chosen by setting $\mathrm{tr}(\dot{K}_{ij}) = \dot{K} = 0$. This yields

$$\frac{1}{(Ae^C)r}\left[\frac{\partial}{\partial r}\left((Ae^C)r\frac{\partial \alpha}{\partial r}\right)\right] + \frac{\partial}{\partial z}\left((Ae^C)r\frac{\partial \alpha}{\partial z}\right)$$

$$= 4\pi\alpha A^2\left[(\rho(1+\epsilon)(W^2 - \frac{1}{2}) + P(W^2 + \frac{1}{2})\right] + \frac{K^j{}_i K^i{}_j}{16\pi}. \quad (6.40)$$

The Hamiltonian constraint is

$$\frac{1}{A^3 e^C r}\left(\frac{\partial^2}{\partial r^2}(Ae^C r) + \frac{\partial^2}{\partial z^2}(Ae^C r)\right) + \frac{1}{A^2}\left(\frac{\partial^2}{\partial r^2}\ln A\frac{\partial^2 \ln A}{\partial z^2}\right)$$

$$= 4\pi\left[(\rho(1+\epsilon)W^2 + P(W^2 - 1) + \frac{K^j{}_i K^i{}_j}{16\pi}\right]$$

$$= 4\pi\left[(D+E)W + (\Gamma-1)E\frac{(W^2-1)}{W} + \frac{K^2}{16\pi}\right]. \quad (6.41)$$

The momentum constraints are satisfied by

$$\nabla^2[A^3 e^C r(K^r{}_r - K^z{}_z)] = 2\left(\frac{\partial P}{\partial r} + \frac{\partial Q}{\partial z}\right), \quad (6.42)$$

$$\nabla^2(A^3 e^C r K^r{}_z) = \frac{\partial P}{\partial z} + \frac{\partial Q}{\partial r}. \quad (6.43)$$

Where now

$$P \equiv A^3 e^C r\left[8\pi\alpha S_r + \frac{1}{2(A^3 e^{3C})r^3}\frac{\partial}{\partial r}((A^3 e^{3C})r^3 K^\phi{}_\phi)\right], \quad (6.44)$$

$$Q \equiv A^3 e^C r\left[8\pi\alpha S_z + \frac{1}{2(A^3 e^{3C})}\frac{\partial}{\partial z}((A^3 e^{3C})K^\phi{}_\phi)\right]. \quad (6.45)$$

The field evolution equations are given as

$$\frac{\partial C}{\partial t} = -\frac{3}{2}\alpha K^\phi{}_\phi - \frac{1}{2}\nabla\cdot\beta + \left[\frac{1}{r} + \frac{\partial C}{\partial r}\right]\beta^r + \left[\frac{\partial C}{\partial z}\right]\beta^z, \quad (6.46)$$

$$\frac{\partial K^\phi{}_\phi}{\partial t} = \beta^r\frac{\partial A}{\partial r} + \beta^z\frac{\partial A}{\partial z} - 8\pi\alpha\left(P + \frac{K^j{}_i K^i{}_j}{16\pi}\right)$$

$$+ \frac{1}{A^2}\left(\frac{\partial^2 \alpha}{\partial r^2} + \frac{\partial^2 \alpha}{\partial z^2}\right) + \frac{\alpha}{A^2}\left(\frac{\partial^2 \ln A}{\partial r^2} + \frac{\partial^2 \ln A}{\partial z^2}\right). \quad (6.47)$$

6.3.1 Axisymmetric hydrodynamics

The hydrodynamic equations are the same as presented in Chapter 3. In this case, however, the metric acceleration terms are

$$\dot{S}_i + \cdots + W\sigma\frac{\partial\alpha}{\partial x^i} + \alpha\sigma\frac{(W^2-1)}{W}\frac{\partial\ln A}{\partial x^i} = 0, \qquad (6.48)$$

where $\sigma \equiv D + \Gamma W$.

To start a calculation an initial distribution of D and E is taken (e.g. two stars in hydrostatic equilibrium). As mentioned above, this is a good case in which to compare the difference between the exact Einstein solution with the conformally flat metric described in the next chapter. This approximation is often invoked as a means to solve the initial value problem, and can even be used to approximate a dynamical system. In this illustration we can similarly solve the initial value problem with a conformally flat metric by simply setting $C = 0$.

The collision was calculated [12] for two cases. First, a collision was computed assuming that the system remained conformally flat, i.e. $C = 0$. The Eqs. (6.46), (6.47) were ignored. The density fluctuates a little as the stars relax around their equilibrium. Overall, however, the density was found to rise as the stars accelerated toward each other. The compression in proper density relative to the proper density of an isolated star was found to rise from 1.0 at the start of the calculation to about 1.7 at the time the stars were near to touching. By this time the square of the spatial part of the four-velocity U^2 had risen to 0.04.

When the full evolution was carried out with $\dot{C} \neq 0$ and $\dot{K}^\phi_\phi \neq 0$, the compression was essentially identical. This increase in central density is comparable to that observed in binary neutron star systems with $U^2 = 0.04$ (cf. [4] and Tables 7.1 and 7.2). This suggests that the compression observed in binary neutron star systems is not an artifact of the conformally flat condition on the metric, but is a real relativistic effect in accelerating systems.

References

[1] J. M. Bardeen and R. V. Wagoner, *Astrophys. J.*, **167**, 359 (1971).

[2] T. Damour, R. S. Hanni, R. Ruffini and J. R. Wilson, *Phys. Rev.*, **17**, 1518 (1978).

[3] J. Hawley, L. Smarr and J. R. Wilson, *Astrophys. J. Suppl. Ser.*, **55**, 211 (1984).

[4] G. J. Mathews and J. R. Wilson, *Phys. Rev.*, **D61**, 127304 (2000).

[5] R. Ruffini and J. R. Wilson, *Phys. Rev.*, **D12**, 2959 (1975).

[6] J. R. Wilson, *Astrophys. J.*, **173**, 431 (1972).

[7] J. R. Wilson, *Astrophys. J.*, **176**, 195 (1972).

[8] J. R. Wilson, *Ann. Phys. NY Acad. Sci.*, **262**, 123 (1975).

[9] J. R. Wilson, in *Proc. First Marcel Grossman Meeting on General Relativity, Trieste, Italy, 1975*, ed. R. Ruffini (North Holland, Amsterdam, 1975), pp. 393–413.

[10] J. R. Wilson, in *Physics and Astrophysics of Neutron Stars and Black Holes (LXV Corso, Soc. Italiana di Fisica, Bologna, Italy), 1978*, ed. R. Giacconi and R. Ruffini (North Holland, Amsterdam, 1978), pp. 643–675.

[11] J. R. Wilson, in *Sources of Gravitational Radiation*, ed. L. Smarr (Cambridge University Press, Cambridge, 1979), p. 423.

[12] J. R. Wilson, *Phys. Rev.*, **D66**, 084015 (2002).

7
Hydrodynamics in three spatial dimensions

Progress in computing full general relativistic hydrodynamics in three spatial dimensions has been slow. The problem is not in the hydrodynamics, but in the solution of the field equations. The equation for extrinsic curvature, K_{ij}, is particularly unstable. For example, at the writing of this book no strong field fully relativistic calculation has computed more than two orbits of a neutron star binary without becoming unstable. Currently, there seems to be some promise, however, in a modified version of the ADM equations based upon a conformal decomposition as originally proposed by Shibata and Nakamura [63] and later reinvented by Baumgarte and Shapiro [7]. At the end of this chapter we briefly summarize this method for completeness. First, however, we summarize a useful alternative which the authors have developed for solving strong field systems which avoids the nonlinearities of the full Einstein equations by reducing the problem to an implementation of constraint equations.

7.1 The conformally flat approximation

For most gravitating systems studied so far (e.g. [5, 23]), only a relatively small amount of energy is emitted by gravitational waves. Even for the merger of two black holes it is expected [1] that only a few tenths of a percent of the rest mass will be radiated away in gravitation. For the case of two neutron stars we would not expect any more radiation to be emitted during the last few orbits than for a two black hole merger, i.e. during the inspiral, the radiated energy per orbit is a minuscule fraction of the energy in orbital motion. Furthermore, an explicit treatment of the radiation reaction is exceedingly difficult [23].

Therefore, we have developed an approximate solution to the ADM equations which omits the cumbersome ADM dynamical equations

181

(cf. Eqs. (1.51) and (1.55)) yet still gives an exact strong field solution to the constraint equations. That is, we simplify the the ADM metric (Eq. (1.41)),

$$ds^2 = -(\alpha^2 - \beta_i\beta^i)dt^2 + 2\beta_i dx^i dt + \gamma_{ij}dx^i dx^j, \qquad (7.1)$$

by requiring that the three-metric be both conformal and flat. Hence, we write,

$$\gamma_{ij} = \phi^4\hat{\gamma}^{ij}, \qquad (7.2)$$

and

$$\hat{\gamma}_{ij} = \delta_{ij}, \qquad (7.3)$$

where the conformal factor ϕ is a positive scalar function describing the ratio between the scale of distance in the curved space relative to the flat space manifold, and δ_{ij} is the Kronecker delta. This is an approximate gauge condition which we refer to as *the conformally flat condition* or *CFC*. This approximation is motivated both by the general observation that gravitational radiation in most systems studied so far is small [1, 23], and by the fact that conformal flatness on each space-like slice considerably simplifies the solution to the field equations. This condition on the metric, however, precludes an explicit manifestation of gravitational radiation. Although gravity waves may be present, they are not manifest without evolving the dynamical equations. Nevertheless, to estimate the signal of emerging gravitational waves we can use a multipole formalism [43, 67, 76]. To estimate their influence on the dynamics, a radiation reaction potential can be added to the hydrodynamics equations to account for the power loss from gravity waves exiting the system.

The implementation of this approximation means that, given a distribution of mass and momentum on some manifold, we first solve the constraint equations of general relativity (GR) at each time in the calculation for a fixed distribution of matter. Then we let the matter and gravitational radiation respond to this geometry. That is, we evolve the hydrodynamic equations to the next time step under an assumption of "instantaneous gravity." However, at each time step we obtain a time symmetric solution to the field equations.

As an alternative to the explicit coupling of emitted gravitational radiation to the hydrodynamic and geometric evolution of the system, the initial evolution of the system (while the gravitational radiation is a small perturbation) can also be approximated by quasi-stable orbits in the absence of energy and momentum loss due to gravitational radiation. One can then, after the fact, compute the expected gradual loss rate of energy and momentum in gravity waves. This latter approach has been often applied [76].

An important advantage of this approach to a solution to the GR equations is that all of the constraint equations reduce to effective flat space elliptic equations which are amenable to standard numerical techniques. Thus, at each time slice we can obtain a numerically valid static solution to the exact GR field equations and information on the hydrodynamic evolution and generation of gravitational radiation. However, the advance from one time slice to the next assumes that the time evolution of the metric can be neglected. Nevertheless, we do incorporate the radiation reaction into the hydrodynamics via a multipole expansion.

The ultimate goal of the binary neutron star problem is to evaluate the gravitational waves emitted near the end of the system's evolution. To obtain the true gravity wave signal one must evolve the binary through many orbits. This evokes the problem of conserving the constraint equations over a long calculational time interval. This is difficult to do with the full ADM equations in three spatial dimensions. Below we describe a conformally flat method of solution that preserves the constraint equations indefinitely. However, the use of a multipole expansion to extract the gravity waves is of doubtful validity in the strong field limit. We are currently developing a method to evolve the gravity waves by a linear perturbation of the full Einstein equations about the conformally flat solution.

7.2 Conformally flat model for binary neutron stars

Coalescing neutron stars are currently of interest for a number of reasons. Several neutron star binaries are known to exist in the Galaxy (e.g. PSR 1913+16 [28], PSR 2303+46 [64], PSR 2127+11C [4], PSR 1534+11 [78]) whose orbits are observed to decay on a time scale of $(1-3) \times 10^8$ years. It has been recognized for some time [16, 17, 19, 60, 61, 69] that the final stages of coalescence of such systems could be copious producers of gravitational radiation. This possibility has received renewed interest with the development of next-generation gravity wave detectors such as cryogenic bar detectors [3], the Caltech-MIT LIGO detector [2], and its European and Japanese counterparts, GEO600, TAMA, and VIRGO (e.g. [15]), for which an event rate due to binary neutron star coalescence out to 200 Mpc could be 3 per year [17, 45, 52]. It has also been proposed that such events could account for the observed event rate and energy requirements of at least some γ-ray bursts [49, 51, 53]. Coalescing neutron stars might even be significant contributors to heavy element nucleosynthesis in the Galaxy [42, 46, 65].

For much of the evolution of a neutron star binary, the system should be amenable to a point source description using post-Newtonian techniques [30, 34, 72]. However, as the stars approach one another the gravitational fields become quite strong and hydrodynamic effects should become significant. Indeed, it is expected that the wave forms could become quite complex as the stars merge. This complexity, however, may be sensitive to various physical properties of the coalescing system [19] such as the neutron star equation of state. Hence, careful modeling is needed which includes both the nonlinear general relativistic effects and a realistic neutron star equation of state. Such calculations can be used as a foundation for extraction of the information contained in the detected gravity waves and as a framework in which to analyze possible γ-ray burst models.

A computation of the hydrodynamic evolution is complicated, however, owing to the inherently three-dimensional character of the orbiting system. To this end several attempts have been made to model the hydrodynamics of coalescence in either a Lagrangian smoothed particle Newtonian approximation [21, 54] or using conventional finite difference methods in the post-Newtonian approximation [44, 48, 49, 55]. It is important to appreciate, however, that as the two neutron stars coalesce the system becomes strongly relativistic, and the validity of Newtonian or post-Newtonian hydrodynamics may be questionable. Here we describe our efforts to model such systems with relativistic hydrodynamics in a conformally flat three-space.

7.2.1 Coordinate system

A number of possible three-space coordinate choices are available to describe binary systems, e.g. polar, bipolar, spherical, cylindrical. Ultimately, we recommend Cartesian x, y, z isotropic coordinates. This is a natural coordinate system for three-dimensional problems, in that no special point or singularity is introduced. It thus avoids problems associated with finite differencing near coordinate singularities. It also has the advantage that the relativistic field equations assume a simpler and more symmetric form.

To see the way in which the CFC allows for a solution to the relativistic field equations, consider Eq. (1.51), the exact dynamical equation [81],

$$\dot{\gamma}_{ij} = -2\alpha K_{ij} + D_i \beta_j + D_j \beta_i, \qquad (7.4)$$

where D_i is the three-space covariant derivative [81], and K_{ab} is the extrinsic curvature describing the deformation of a figure as it is carried forward by one unit in proper time in a direction normal to a hypersurface.

Equation (7.4) is well approximated by a conformal representation (7.2) *only if* the trace-free part of the right-hand side vanishes. Thus, a spatially flat three-metric requires,

$$2\alpha K_{ij} = (D_i\beta_j + D_j\beta_i - \frac{2}{3}\gamma_{ij}D_k\beta^k), \tag{7.5}$$

where we have employed the maximal slicing condition, $\text{tr}(K_{ij}) = K^i{}_i = 0$ as a gauge choice.

We use Eq. (7.5) to determine the extrinsic curvature. A convenient consequence of this is that any geometry which is initially conformally flat will remain conformally flat to the extent that energy in gravitational radiation is unimportant. Equation (7.5) allows us to derive constraint equations for the lapse function and conformal factor as described in the next section.

As a final condition, we take the coordinate system (shift vector) to be rotating in such a way as to minimize the matter motion in the coordinate grid. We also add a separate grid velocity along the line between centers of the stars. This condition enhances the accuracy of the computation of the hydrodynamic evolution. However, this is a nontrivial condition to impose in curved spacetime which we achieve by boundary conditions on β^i along the direction of orbital motion as described below. The additional grid velocity V_g^i along the direction between the centers of the stars highlights the fact that the shift vector and grid velocity can be two independent coordinate degrees of freedom. Nevertheless, it is important to remember that all relevant forces are computed first in nonrotating Eulerian observer coordinates (cf. Figure 1.2) which are then transformed to update the matter fields in a rotating grid.

7.2.2 Hamiltonian constraint

We begin with the Hamiltonian constraint equation [81] (Eq. (1.59)). We use the forms of equations as given by Evans [23]. We show here that the Hamiltonian constraint and the maximal slicing condition ($\text{tr}(K_{ij}) = K^i{}_i \equiv K = \dot{K} = 0$) can be combined so as to form elliptic equations for both the conformal factor ϕ and the product $(\alpha\phi)$.

The Hamiltonian constraint equation (Eq. (1.59)) can be written,

$$R = 16\pi\rho_H + K_{ij}K^{ij} - K^2, \tag{7.6}$$

where R is the Ricci scalar curvature (Eq. (1.6)), and ρ_H is the Hamiltonian density (1.33).

The conformal scaling of the three-metric, Eq. (7.2), defines a conformal metric and manifold $(\hat{\gamma}, \hat{M})$ related to the physical metric and manifold (γ, M) (see [23, 81]). Covariant derivatives D_i and $\hat{D}_i$ on M and $\hat{M}$ can be related by calculating the transformation of the Christoffel connections,

$$\Gamma^i_{jk} = \hat{\Gamma}^i_{jk} + 2\phi^{-1}\left[\delta^i_j \hat{D}_k\phi + \delta^i_k \hat{D}_j\phi - \hat{\gamma}_{jk}\hat{\gamma}^{il}\hat{D}_l\phi\right]. \tag{7.7}$$

With this, the transformation of the Ricci scalar curvature is

$$R = \phi^{-4}\hat{R} - 8\phi^{-5}\hat{\Delta}\phi, \tag{7.8}$$

where $R = R(\gamma)$, $\hat{R} = \hat{R}(\hat{\gamma})$, and $\hat{\Delta} = \hat{\gamma}^{ij}\hat{D}_i\hat{D}_j$. As mentioned in Section 7.2.1, we choose a conformally flat metric, $\hat{\gamma}^{ij} = \delta^{ij}$, for which, $\hat{\Gamma}^i_{jk} \to 0$, $\hat{D}_i \to \nabla$, $\hat{R} \to 0$, and $\hat{\Delta} \to \nabla^2$, the flat space Laplacian. Solving Eq. (7.8) for $\hat{\Delta}\phi$, and combining with the Hamiltonian constraint gives the desired form for an elliptic equation for ϕ,

$$\nabla^2\phi = -2\pi\phi^5\left[\rho_H + \frac{K_{ij}K^{ij}}{16\pi}\right]. \tag{7.9}$$

In order to put this constraint equation into a form which is useful for solution along with the hydrodynamic variables, we must introduce conformal scalings for the source terms.

In terms of the usual equation of state index Γ, Eq. (1.33) becomes

$$\rho_H = \rho W^2 + \rho\epsilon W\left[\Gamma W - \frac{(\Gamma - 1)}{W}\right]. \tag{7.10}$$

For the hydrodynamic Lorentz-contracted coordinate density, $D = \rho W$, and energy, $E = \rho\epsilon W$, we introduce the following conformal scalings,

$$D = \phi^{-6}\hat{D}, \tag{7.11}$$

$$E = \phi^{-6\Gamma}\hat{E}. \tag{7.12}$$

The reasons for these choices will be clear when we consider the hydrodynamic equations given in Section 7.3.

The extrinsic curvature is scaled by

$$K^{ij} = \phi^{-10}\hat{K}^{ij}, \tag{7.13}$$

which gives

$$K_{ij} = \phi^{-2}\hat{K}_{ij}. \tag{7.14}$$

With the introduction of these scalings, the Hamiltonian constraint can be written in a familiar Poisson form,

$$\nabla^2\phi = -4\pi\rho_1, \tag{7.15}$$

in which the source term can be identified in terms of physical hydrodynamic variables by transforming the conformal scalings in Eqs. (7.11)–
(7.14),

$$\rho_1 = \frac{\phi^5}{2}\left[DW + E\left(\Gamma W - \frac{(\Gamma-1)}{W}\right) + \frac{1}{16\pi}K_{ij}K^{ij}\right].\qquad(7.16)$$

7.2.3 Lapse function

We also use the Hamiltonian constraint together with the maximal slicing
condition $\mathrm{tr}(K_{ij}) \equiv K = \dot{K} = 0$ to obtain an elliptic equation for the
product of the lapse function and conformal factor $(\alpha\phi)$. We begin with
the identities,

$$D_iD^i\alpha \equiv \Delta\alpha = \Delta[\phi^{-1}(\alpha\phi)] = D_iD^i[\phi^{-1}(\alpha\phi)]\qquad(7.17)$$
$$= \phi^{-1}\Delta(\alpha\phi) - 2\phi^{-6}\hat{\gamma}^{ij}\hat{D}_i\phi\hat{D}_j(\alpha\phi) + \alpha\phi\Delta(\phi^{-1}).\quad(7.18)$$

Now in our conformally flat metric one can write for any scalar function,
and in particular for the quantity $(\alpha\phi)$,

$$\Delta(\alpha\phi) = \phi^{-4}\hat{\Delta}(\alpha\phi) + 2\phi^{-5}\hat{\gamma}^{ij}(\hat{D}_i\phi)(\hat{D}_j(\alpha\phi)).\qquad(7.19)$$

Substituting this into Eq. (7.18) gives

$$\Delta\alpha = \phi^{-5}\hat{\Delta}(\alpha\phi) + \alpha\phi\Delta(\phi^{-1}).\qquad(7.20)$$

Now from the transformation properties of the Ricci curvature scalar
(7.8), Eq. (7.20) can be rearranged as

$$\hat{\Delta}(\alpha\phi) = \phi^5\Delta\alpha + \frac{1}{8}\alpha\phi^5\left[\hat{R}\phi^{-4} - R\right].\qquad(7.21)$$

Setting $K = \dot{K} = 0$ in Eq. (1.53) can be used to find an expression for
$\Delta\alpha$. Then rewriting the Hamiltonian constraint (7.6), to include the CFC
and maximal slicing conditions, leads ultimately to a flat space elliptic
equation in $(\alpha\phi)$. In Poisson-like form this is

$$\nabla^2(\alpha\phi) = 4\pi\rho_2,\qquad(7.22)$$

with the source term written in terms of hydrodynamic variables as

$$\rho_2 = \frac{\alpha\phi^5}{2}\left[\frac{D(3W^2-2) + E[3\Gamma(W^2+1)-5]}{W} + \frac{7}{16\pi}K_{ij}K^{ij}\right].\qquad(7.23)$$

A solution of Eq. (7.22) determines the lapse function after Eq. (7.15) is
used to determine the conformal factor.

7.2.4 Momentum constraint

With the lapse function and conformal factor determined from the Hamiltonian constraint and maximal slicing condition, we then use the momentum constraints to find the shift vector.

The momentum constraints (Eq. (1.60)) have the form [23],

$$D_i(K^{ij} - \gamma^{ij}K) = 8\pi s^j. \tag{7.24}$$

Where D_j is the three-space covariant derivative [81], and s^i is the contravariant *three*-momentum density. We write this in a small character to avoid confusing it with the spatial components of the four-momentum which is what is solved in the hydrodynamic equations (cf. Section 7.3). It is easy to become confused on this point, as we have shown [26, 40].

In this approximation scheme, we impose maximal slicing $(\mathrm{tr}(K^{ij}) = 0)$ and demand that the spatial three-metric γ_{ij} be conformally flat. The second term on the left-hand side then vanishes and we have,

$$D_i K^{ij} = 8\pi s^j. \tag{7.25}$$

Equation (7.24) ultimately reduces to a Poisson-like equation for the shift vector,

$$\nabla^2 \beta^i = 4\pi \rho_\beta^i - \frac{\partial}{\partial x^i}\left(\frac{1}{3}\nabla \cdot \beta\right). \tag{7.26}$$

Thus, by introducing a decomposition of β^i into

$$\beta^i = B^i - \frac{1}{4}\partial_i \chi, \tag{7.27}$$

the following two elliptic equations result:

$$\nabla^2 \chi = \frac{4}{3}\nabla \cdot \beta, \tag{7.28}$$

$$\nabla^2 B^i = \rho_\beta^i, \tag{7.29}$$

where

$$\rho_\beta^i = 4\alpha\phi^4 S_i + \frac{1}{4\pi\xi}\frac{\partial\xi}{\partial x^j}\left(\frac{\partial\beta^i}{\partial x^j} + \frac{\partial\beta^j}{\partial x^i} - \frac{2}{3}\delta_{ij}\frac{\partial\beta^k}{\partial x^k}\right), \tag{7.30}$$

where $\xi \equiv \alpha/\phi^6$ and now S_i are spatial components of the covariant four-momentum density determined from the hydrodynamics. Equations (7.28)–(7.30) can then be used to determine the components of the shift vector.

The meaning of orbital angular velocity becomes obscured in curved spacetime. Nevertheless, we wish to identify an orbital frequency ω with which to identify an orbital grid velocity. To do this we introduce ω as a Lagrange multiplier which minimizes the matter motion with respect to the coordinate system. Confining orbital motion to the x, y plane, we determine the coordinate rotation frequency ω at each time step from the weighted average of the matter four-velocity and the frame-drag shift vector,

$$\omega = \frac{\int dV (D + \Gamma E) \left[\frac{\alpha(xU_y - yU_x)}{(1 + U^\cdot/\phi^\cdot)} - \phi^4(xG_y - yG_x) \right]}{\int dV (D + \Gamma E)(x^2 + y^2)}. \tag{7.31}$$

This rotation is then subtracted as a grid velocity $\vec{V}_g = \omega \times \vec{R}$ and added as a coordinate rotation. This maintains the centering of the stars along $y = 0$.

The fact that the constraint conditions on Eqs. (7.15), (7.22), and (7.26) can be written in the form of flat space Poisson equations allows for these variables to be solved by fast numerical techniques as discussed below. However, their solution requires that boundary values for these variables be specified at distances relatively close to the neutron stars. Our method of determining the boundary values is described in Section 7.4.

7.2.5 Reliability of the conformally flat condition

Two questions are relevant when considering whether to take advantage of the CFC. One is the validity of this metric choice for the initial value problem, and the other is the effect on the system of the "hidden" gravitational radiation in the physical data.

Regarding the validity of the CFC one has a great deal of freedom in choosing coordinates and initial conditions as long as the initial space is Riemannian and the metric coefficients satisfy the constraint equations of general relativity [43]. Indeed, we have shown in [6] that exact solutions for the CFC metric coefficients can be obtained by imposing the ADM Hamiltonian and momentum constraint conditions. Nevertheless, in three dimensions a physical space is conformally flat if and only if the Cotton–York tensor vanishes [22, 31],

$$C^{ij} = 2\epsilon^{ikl} \left(R^j{}_k - \frac{1}{4} \delta^j{}_k R \right)_{;l}, \tag{7.32}$$

where $R^j{}_k$ is the Ricci tensor and R is the Ricci scalar for the three-space.

Equation (7.32) vanishes by fiat for the three-space metric we have chosen. However, conformally flat solutions for physical problems have only been proven [22, 31] for spaces of special symmetry (e.g. constant curvature, spherical symmetry, time symmetry, Robertson–Walker, etc. [31]). Hence, the invocation of the CFC here and in other applications is an assumption. That is, it is a valid solution to the Einstein constraint equations, but does not necessarily describe a physical configuration to which two neutron stars will evolve. Nevertheless, this is a valid approximation as long as the nonconformal contributions from the $\dot{\gamma}_{ij}$ and $\dot{K}_{ij}$ equations in the exact two neutron star problem remain small. Indeed, numerical tests for an axisymmetric rotating neutron star [18] have indicated that conformal flatness is a good approximation when it can be tested.

As a related illustration, consider the Kerr solution for a rotating black hole. It is well known that the Kerr metric is not conformally flat. The close binaries we study have specific angular momentum only slightly greater than that of an extreme Kerr black hole. Also, they ultimately merge and collapse to a single Kerr black hole. Hence, an analysis of the Cotton–York tensor for a Kerr black hole is another indicator of the degree to which conformal flatness is a valid approximation for neutron-star binaries.

Figure 7.1 from [41] gives the dimensionless scaled Cotton–York parameter $C^{\theta\phi}m^3$ for a maximally rotating Kerr black hole as a function

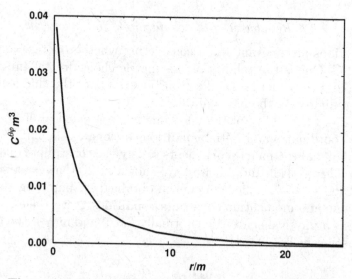

Fig. 7.1. The scaled Cotton–York tensor component $C^{\theta\phi}m^3$ as a function of proper radius r/m for a maximally rotating $a = m$ Kerr black hole. This quantity is a measure of the deviation from conformal flatness.

of proper distance. For illustration, consider the decrease of this quantity as one moves away from the horizon at $m = r$ as a measure of the rate at which the metric becomes conformally flat. The maximally rotating $(a = m)$ black hole of this example, however, is an extreme example of compactness and angular velocity relative to any orbiting neutron stars.

It can be seen in Figure 7.1 that, even for this extreme case, the dimensionless tensor coefficient $C^{\theta\phi}m^3$ diminishes rapidly away from the black hole. At the separation of interest for binary neutron stars approaching their final orbits ($r/m \sim 10$ where m is the total binary mass and r the half separation between stars), this coefficient has already diminished to approximately a few 10^{-3} of the value at the event horizon ($r/m \sim 1$). Thus, the effect of either star on its companion is probably well approximated by conformal flatness. Regarding the interiors of the neutron stars themselves, in most cases the stars are rotating so slowly (even when corotating) that the deviation from conformal flatness is probably negligible. Thus, it seems plausible that conformal flatness is a reasonable approximation for most physical aspects involving the spatial three-metric of binary neutron star systems.

The next issue concerns the "hidden" radiation in the physical data. To address this we decompose the extrinsic curvature into longitudinal K_L^{ij} and transverse K_T^{ij} components as proposed by York [80],

$$K^{ij} = K_L^{ij} + K_T^{ij}. \tag{7.33}$$

By definition the transverse part obeys

$$D_i K_T^{ij} = 0, \tag{7.34}$$

where D_i are covariant derivatives. The longitudinal part can be derived from a properly symmetrized vector potential. We find

$$D_i K_L^{ij} = 8\pi S^i, \tag{7.35}$$

where S^i are spatial components of the contravariant four-momentum density.

The product $K_T^{ij} K_{Tij}$ is a measure of the hidden radiation energy density. To find K_T^{ij} from our numerical calculations, we first find K_{ij} by choosing maximal slicing ($\text{tr}(K_{ij}) = 0$) and requiring that the trace-free part of the $\dot{\gamma}_{ij}$ equation vanish. This gives [41, 76]

$$2\alpha K_{ij} = (D_i\beta_j + D_j\beta_i - \frac{2}{3}\phi^{-4}\delta_{ij}D_k\beta^k). \tag{7.36}$$

We then determine K_L^{ij} from the equilibrium momentum density using Eq. (7.35) and subtract K_L^{ij} from K^{ij}.

We find [41] that this measure of the "hidden" gravitational radiation energy density is a small fraction of the total gravitational mass energy of the system,

$$\int K_T^{ij} K_{Tij} \frac{dV}{8\pi} \approx 2 \times 10^{-5} \, M_G. \tag{7.37}$$

Hence, we conclude that the CFC is probably a good approximation to the initial data.

This should be an excellent approximation for the determination of stellar structure and stability. However, an unknown uncertainty enters if one attempts to reconstruct the time evolution of the system (e.g. the gravitational waveform) from this sequence of quasi-static initial conditions. At present we make this connection approximately via a multipole expansion [68] for the gravitational radiation as described below in Section 7.3.2.

7.2.6 Other checks on the conformally flat condition

Since conformal flatness implies no transverse traceless part of γ_{ij} it can minimize the amount of gravitational radiation apparent in the initial configuration. However, in general the physical data still contain a small amount of preexisting gravitational radiation. This has been clearly demonstrated in numerical calculations of axisymmetric black hole collisions [1]. In exact numerical simulations, the gravitational radiation appears as the time derivatives of the spatial three-metric ($\dot\gamma_{ij}$) and its conjugate (the extrinsic curvature $\dot K_{ij}$) are evolved. The immediate evolution of the fields from conformally flat initial data is characterized by the development of a weak gravity wave exiting the system.

An estimate of the radiation content of initial data slices for axisymmetric black hole collisions has been made by Abrahams [1]. Even for high values of momentum, the initial slice radiation is always less than about 10% of the maximum possible radiation energy (as estimated from the area theorem).

Regarding the reliability of the CFC as an approach to the initial value problem a study by Cook *et al.* [18] has shown that an axially symmetric CFC approximation is quite good when computed physical observables are compared with the exact results for axisymmetric extremely rapidly rotating neutron stars. This is the simplest system for which an exact metric begins to differ from a CFC metric.

In another example of an axisymmetric application of the CFC the head-on collision of two neutron stars has been studied using both an exact solution of the field and hydrodynamic equations [74] (cf. Chapter 6) and using using the CFC. As shown in Section 6.3.1, the conformally flat and exact solutions are essentially identical to good precision. This is an important test in that it confirms that the central density of the stars increases as they accelerate prior to collision.

7.3 Relativistic hydrodynamics

In the CFC, the hydrodynamic equations of motion are simplified over the more general equations of Chapter 3. In particular, the γ terms simplify since we now have

$$\gamma = \phi^6. \tag{7.38}$$

This gives a new expression for the Lorentz-like factor,

$$W = \alpha U^t = \left[1 + \frac{\sum U_i^2}{\phi^4}\right]^{1/2}. \tag{7.39}$$

The continuity equation takes the form,

$$\frac{\partial D}{\partial t} = -6D\frac{\partial \log \phi}{\partial t} - \frac{1}{\phi^6}\frac{\partial}{\partial x^j}(\phi^6 DV^j). \tag{7.40}$$

The energy equation becomes

$$\frac{\partial E}{\partial t} = -6\Gamma E\frac{\partial \log \phi}{\partial t} - \frac{1}{\phi^6}\frac{\partial}{\partial x^j}(\phi^6 EV^j)$$
$$- P\left[\frac{\partial W}{\partial t} + \frac{1}{\phi^6}\frac{\partial}{\partial x^j}(\phi^6 WV^j)\right], \tag{7.41}$$

while the momentum equation becomes

$$\frac{\partial S_i}{\partial t} = -6S_i\frac{\partial \log \phi}{\partial t} - \frac{1}{\phi^6}\frac{\partial}{\partial x^j}(\phi^6 S_i V^j) - \alpha\frac{\partial P}{\partial x^i}$$
$$+ 2\alpha(D + \Gamma E)(W - \frac{1}{W})\frac{\partial \log \phi}{\partial x^i} + S_j\frac{\partial \beta^j}{\partial x^i}$$
$$- W(D + \Gamma E)\frac{\partial \alpha}{\partial x^i} - \alpha W(D + \Gamma E)\frac{\partial \chi}{\partial x^i}. \tag{7.42}$$

In Eq. (7.42) we have now added the radiation reaction potential, χ. The evaluation of this quantity is described in Section 7.3.2. The repeated occurrence of the ϕ^6 factors simply maintains the proper volume (proper volume is $\phi^6(dx^i)^3$). This is the reason for the choice of conformal scalings introduced in Eqs. (7.11) and (7.12). By making this transformation when solving the advection equations, we can use the methods described in Section 2.4.9. Also, when the metric is updated at the end of the cycle (cf. Section 3.3) we can preserve $\hat{D}\hat{E}$, and $\hat{S}_i$ as ϕ is changed.

7.3.1 Equation of state

For the orbital calculations presented here we use the zero temperature, zero neutrino chemical potential equation of state from [76]. While the orbital calculations of concern here should only involve zero temperature, there is some small shock heating of the stars as they adjust to changing conditions on the grid. Thus, we augment this equation of state with a thermal component (taken to behave as a $\Gamma = 5/3$ gas) in order to follow the dynamic evolution equations. Thus, we write

$$P = P_0(\rho) + \frac{2}{3}\rho(\epsilon - \epsilon_0(\rho)), \tag{7.43}$$

where P_0 and ϵ_0 are the zero temperature pressures and energies.

7.3.2 Gravitational radiation

In general it is possible to express the emission of gravitational radiation in terms of an "exact" expansion [68] of multipole moments of the effective stress energy tensor, including corrections for the so-called "slow motion" approximation. It is important to appreciate that these formulas apply to strong field sources as well as to weak field sources as long as the relevant components of the effective stress energy tensor can be identified. Since here we are only concerned with orbital motion of equal mass binaries, the multipole expansions reduce to only a few nonzero terms. These can be evaluated and tested for convergence of the expansion (cf. [76]).

In any coordinate system (such as the one we are using here) in which the gravity waves far from the source can be characterized as linear metric perturbations propagating on a flat background, the transverse traceless part of the metric perturbation characterizes the radiation completely. This metric perturbation can be expressed [68] in terms of the mass multipole (I^{lm}) and current multipole moments (S^{lm}) as

$$h_{jk}^{TT} = \sum_{l=2}^{\infty} \sum_{m=-l}^{l} \left[r^{-1} \, {}^{(l)}I^{lm}(t-r)T_{jk}^{E2,lm} + r^{-1} \, {}^{(l)}S^{lm}(t-r)T_{jk}^{B2,lm} \right], \tag{7.44}$$

where the superscript TT denotes the transverse traceless part of the metric perturbation and the notation ${}^{(l)}I^{lm}$ and ${}^{(l)}S^{lm}$ denotes the lth time derivative of the respective moments. The quantities $T_{jk}^{E2,lm}$ and $T_{jk}^{B2,lm}$ are *pure spin spherical harmonics* as defined in [68].

From this, the general expression for energy loss is

$$\frac{dE}{dt} = \frac{1}{32\pi} \sum_{l=2}^{\infty} \sum_{m=-l}^{l} \langle |^{(l+1)}I^{lm}|^2 + |^{(l+1)}S^{lm}|^2 \rangle, \qquad (7.45)$$

where the brackets denote averages over several wavelengths. Angular momentum loss can similarly be written

$$\frac{dJ}{dt} = \frac{i}{32\pi} \sum_{l=2}^{\infty} \sum_{m=-l}^{l} \langle^{(l)}I^{lm*}m^{(l+1)}I^{lm}\rangle + \langle^{(l)}S^{lm*}m^{(l+1)}S^{lm}\rangle, \quad (7.46)$$

where Eq. (7.46) assumes an alignment of the angular momentum vector with the z-axis.

The radiation reaction potential χ for Eq. (7.42) can be written

$$\chi = \frac{1}{32\pi} \sum_{l=2}^{\infty} \sum_{m=-l}^{l} x_i x_j \langle |^{(l+1)}I^{lm}|^2 + |^{(l+1)}S^{lm}|^2 \rangle. \qquad (7.47)$$

The problem then reduces to the identification of the relevant mass and current moments in the chosen coordinates. For a coordinate system which asymptotically becomes Minkowski, one can define a quantity

$$\bar{h}^{\alpha\beta} \equiv -(-g)^{1/2}g^{\alpha\beta} + \eta^{\alpha\beta}, \qquad (7.48)$$

where g is the determinant of the metric and $\eta^{\alpha\beta}$ is the Minkowski metric tensor. If $\bar{h}^{\alpha\beta}$ satisfies the de Donder gauge condition

$$\bar{h}^{\alpha\beta}{}_{,\beta} = 0, \qquad (7.49)$$

then the Einstein field equations take the form

$$\Box \bar{h}^{\alpha\beta} = -16\pi \tau^{\alpha\beta}, \qquad (7.50)$$

where $\tau^{\alpha\beta}$ is the "effective stress energy tensor" [68].

As long as the de Donder condition is valid, Eq. (7.50) can be inverted (using the flat space outgoing Green's function) and the Green's function expanded in terms of vacuum basis functions. The resultant expression can then be reduced [68] to provide expansions for the desired mass and current moments:

$$
I^{lm} = \frac{16\pi}{(2l+1)!!} \left(\frac{(l+1)(l+2)}{2(l-1)l} \right)^{1/2} \int \tau_{00} Y^{lm*} r^l d^3x
$$

$$
+ \sum_{k=0}^{\infty} \frac{16\pi}{2^k k! (2l+2k+1)!!} (\partial_t)^{2k} \int \tau_{pq} r^{l+2k}
$$

$$
\times \left[\frac{(2l+2k+1)}{2(k+1)} \left(\frac{(l+1)(l+2)}{2(2l-1)(2l+1)} \right)^{1/2} T_{pq}^{2\ l-2,lm*} \right.
$$

$$
+ \left(\frac{3(l-1)(l+2)}{(2l-1)(2l+3)} \right)^{1/2} T_{pq}^{2\ l,lm*} + \frac{2k}{2l+2k+3}
$$

$$
\left. \times \left(\frac{l(l-1)}{2(2l+1)(2l+3)} \right)^{1/2} T_{pq}^{2\ l+2,lm*} \right] d^3x, \tag{7.51}
$$

and

$$
S^{lm} = \frac{-32\pi}{(2l+1)!!} \left(\frac{(l+2)(2l+1)}{2(l-1)(l+1)} \right)^{1/2}
$$

$$
\times \int \epsilon_{jpq} x_p (-\tau_{0q}) Y_j^{l-1,lm*} r^{l-1} d^3x
$$

$$
+ \sum_{k=0}^{\infty} \frac{16\pi i}{2^k k! (2l+2k+1)!!} (\partial_t)^{2k+1} \int \tau_{pq} r^{l+2k+1}
$$

$$
\times \left[\frac{1}{2(k+1)} \left(\frac{l+2}{2l+1} \right)^{1/2} T_{pq}^{2\ l-1,lm*} \right.
$$

$$
\left. + \frac{1}{2l+2k+3} \left(\frac{l-1}{2l+1} \right)^{1/2} T_{pq}^{2\ l+1,lm*} \right] d^3x, \tag{7.52}
$$

where the Y^{lm*} are the usual spherical harmonics, and $T_{pq}^{2\ l,lm*}$ are the pure orbital tensor harmonics as defined in [68]. The first integral in Eqs. (7.51) and (7.52) is the usual spherical harmonic expansion. At the $l=2$ level, Eq. (7.51) reduces to the well known quadrupole approximation. The second integral in Eqs. (7.51) and (7.52) is the correction to the slow motion approximation, which is non-negligible in the present application, i.e. v/c 0.1.

To evaluate the time derivatives of the mass and current multipole moments one can make use of the rotation properties of spherical tensors whereby, rotations can be generated in terms of the Wigner D matrices,

$$
I^{lm} = D_{mm'}^l I_0^{lm'}, \qquad S^{lm} = D_{mm'}^l S_0^{lm'}, \tag{7.53}
$$

where $I_0^{lm'}$ and $S_0^{lm'}$ are evaluated in the rotating frame. For stable orbits (neglecting gravitational radiation) and hydrostatic stars, these are time independent quantities.

The main contribution to the time derivatives is that due to orbital motion. Evaluation of the orbital motion reduces to derivatives of the $D_{mm'}^l$ which for Cartesian coordinates have a simple $\sim\cos(m\omega t)$ dependence.

The problem with evaluating Eqs. (7.51) and (7.52) is that the multipole moments are only defined in the de Donder gauge and not for our conformally flat coordinates. Furthermore, even if the transformation to our coordinates were straightforward (which it is not) the effective stress energy tensor would not be known.

Fortunately, however, a transformation to de Donder coordinates is not necessary. It is only necessary that the moments of the metric coordinates be defined in a coordinate system which, like a de Donder coordinate system, is asymptotically Cartesian and mass centered (ACMC). In [68] it is proven that in such coordinate systems the covariant metric components are time independent and expandable into a spherical harmonic $(1/r)$ structure in terms of the same moments (i.e. Eqs. (7.51) and (7.52)) relevant to the radiation field. Furthermore, these multipole moments are invariant under transformations between two ACMC coordinate systems. From these expansions we can deduce the source for the slow motion moments to be used in the equations for the radiation field ((7.45)–(7.46)). For example, the spatial three-metric must obey [68]:

$$\gamma_{ij} = \delta_{ij} + \sum_{l=0}^{N} \frac{1}{r^{l+1}} \left[\frac{(2l-1)!!}{2} \left(\frac{2(l-1)l}{(l+1)(l+2)} \right)^{1/2} \right.$$

$$\left. \times \sum_{m=-l}^{l} I^{lm} Y^{lm} + (l-1 \ pole) + \cdots + (0 \ pole) \right]. \quad (7.54)$$

However, the spatial three-metric (Eq. (7.2)) can also be expanded as the fourth power of a multipole expansion of the flat space Poisson equation for ϕ (Eq. (7.15)),

$$\gamma_{ij} = \phi^4 \delta_{ij}$$

$$= \left[1 + \sum_{l=0}^{\infty} \sum_{m=-l}^{l} \frac{4\pi}{(2l+1)} q^{lm} Y^{lm} r^{-(l+1)} \right]^4 \delta_{ij}, \quad (7.55)$$

where

$$q^{lm} = \int d^3x \rho_1(x) r^l Y^{lm*}, \quad (7.56)$$

and ρ_1 is the source term for ϕ (Eq. (7.16)). If we collect the dominant linear terms in Eqs. (7.54) and (7.55) according to the recipe given in [68], then we can identify the relation between the source ρ_1 for the conformal factor elliptic equation (7.15) and the mass multipole moments, i.e.

$$I^{lm} = \frac{32\pi}{(2l+1)!!}\left(\frac{(l+1)(l+2)}{2(l-1)l}\right)^{1/2} q^{lm}. \tag{7.57}$$

This identification also reduces to the correct Newtonian limit. As can be seen from Eq. (7.16), $\rho_1 \to \rho/2$, where ρ is the Newtonian matter density, so that $\tau_{0d} \to \rho$ as required.

The contribution from the current moments is expected to be small, as is the slow motion correction. Therefore, we are mainly concerned with estimating the magnitude of those contributions. To the accuracy desired, we identify the source for the current moments S^{lm} and the slow motion corrections with the Newtonian-like counterparts, i.e. we set $\tau_{0j} = T_{0j}$, $\tau_{ij} = T_{ij}$. We have computed [76] terms out to ω^{10}, which includes mass multipoles out to $l = 4$, current multipoles out to $l = 3$ and the leading correction for the slow motion correction. For quasi-static binary neutron star orbits, as expected, the quadrupole term dominates. The next largest term is the slow motion correction which contributes only a few percent to the gravitational radiation and tends to decrease the loss rate.

7.3.3 Solution of elliptic equations

As shown above, the solution of the field equations reduces to evaluating numerically a set of elliptic equations. After finite differencing, the elliptic equations are reduced to a matrix equation,

$$M \cdot x = b, \tag{7.58}$$

where M is a sparse matrix, x is a vector representing the relevant field variable at each zone, and b is derived from the source terms. This equation can then be solved using any one of a number of fast matrix inversion techniques. The algorithms available include variants of the conjugate gradient method or successive over relaxation. We have found that the *diagonal scaling conjugate gradient method* makes best use of computer resources for this problem, although the *incomplete Coleski* converges faster if one has the computer memory available. Probably, the most widely used method currently is that of multi-grid successive over relaxation.

As another important reminder, when solving the elliptic equation for ϕ, the coordinate density D should be adjusted so as to preserve the conformal scalings, Eqs. (7.11) and (7.12). That is, $\hat{D} = \phi^6 D$ should be kept constant. As the metric is updated, this preserves baryon number. Also, the coordinate energy density should be changed to preserve $\phi^{6\Gamma} E$ which conserves entropy, and the momentum density should be changed to preserve $\phi^6 S^i$, which maintains the momentum.

7.3.4 Extracting physical observables

In the CFC, the gravitational mass can be obtained from the asymptotic behavior of $\phi \to 1 + (GM/2r)$ (cf. Eq. (7.55)). The angular momentum, however, is more difficult to define. We estimate this from an integral over the spacetime components of the stress energy tensor [43] neglecting angular momentum in the radiation field,

$$J^{ij} = \int \left(T^{i0} x^j - T^{j0} x^i \right) dV. \tag{7.59}$$

Aligning the z-axis with the angular momentum vector then gives

$$J = \int (x S^y - y S^x) dV. \tag{7.60}$$

7.4 Boundary conditions

As noted above, these choices for the metric and slicing condition lead to a form for the Hamiltonian and momentum constraints in terms of flat space elliptic equations, i.e. Eqs. (7.15), (7.22), and (7.26), for the metric variables ϕ, $(\alpha\phi)$, and β^i. A solution to these elliptic equations, however, requires that we specify values for ϕ, $(\alpha\phi)$, and β^i along the outer boundaries of the grid. For a Poisson-like equation, the field variables could be specified by integrating the source function over the interior, e.g.

$$\phi(x) = \int \frac{\rho_1(x')}{|x - x'|} d^3 x'. \tag{7.61}$$

However, the evaluation of this integral for each point along the boundaries is computationally slow. In principle, an expansion of the source function in spherical harmonics $Y^{lm}(\theta, \phi)$ could be applied to obtain the field variables along the boundaries.

Since the fields obey Poisson-like equations, they are additive in the vacuum region. An expansion in spherical harmonics can then be made in the frame of each star $i = 1, 2$,

$$\tilde{\phi}_i(x) = \sum_{l}^{\infty} \sum_{m=-l}^{l} \frac{4\pi}{2l+1} r^{-(l+1)} q_i^{lm} Y^{lm}(\theta, \phi), \tag{7.62}$$

where r is the distance from the center of the star, and the moments q_i^{lm} are evaluated in the portion of the grid associated with star i. Obviously, this distinction breaks down as the stars come into contact and one must revert at some point to evaluating a single set of moments over the whole grid.

As long as the two stars are distinct, the field on the boundaries x can be given by

$$\phi(x) = 1 + \tilde{\phi}_1(\vec{x} - \vec{x}_1) + \tilde{\phi}_2(\vec{x} - \vec{x}_2), \qquad (7.63)$$

where $\vec{x}_1$ and $\vec{x}_2$ are the position coordinates of the two stars. In practice the harmonic expansion for each star converges rapidly and a truncation of the expansion at $l = 4$ is sufficient.

We note that the expansion of the three-metric (Eq. (7.54)) requires that the asymptotic form for ϕ obey

$$\phi \to 1 + \frac{m_G}{2r}. \qquad (7.64)$$

Similarly, from the ACMC expansion for g_{00} [68] the lapse function must approach

$$\alpha \to 1 - \frac{m_G}{r}, \qquad (7.65)$$

in order that the time coordinate become proper time as $r \to \infty$. The Poisson equation (7.22) for $(\alpha\phi)$ can also be expanded in spherical harmonics (e.g. Eq. (7.62)) yielding

$$(\alpha\phi) \to (\alpha\phi)_\infty - \frac{m_{\alpha\phi}}{2r}, \qquad (7.66)$$

where $m_{\alpha\phi}$ is the volume integral over twice the source ρ_2. Since in general $m_{\alpha\phi} \neq m_G$, we choose the boundary condition,

$$(\alpha\phi)_\infty = \frac{m_{\alpha\phi}}{m_G}, \qquad (7.67)$$

to guarantee that Eq. (7.65) is satisfied. Typically, $(\alpha\phi)_\infty \approx 0.98$.

In the computation of the boundary conditions, one must impose a spherical cut-off in the matter distributions at a radius equal to the largest sphere that fits within the cubic grid. This avoids the possibility of a spurious hexadecapole moment associated with the cubic grid employed in the calculation. For matter terms this is a reasonable truncation for the calculations presented here, since only a negligible amount of matter appears beyond the surface of the neutron stars. However, the $K_{ij}K^{ij}$

terms in Eqs. (7.15) and (7.22) contribute beyond the matter boundary. Also, the shift vector elliptic equations, (7.28) and (7.29), involve a source which extends beyond the source boundary.

Regarding the $K_{ij}K^{ij}$ terms we note that these terms are small. For example, the contribution to the gravitational mass from an integration over the interior source function is only 0.0001 $M_\odot$. Furthermore, the asymptotic form for $K_{ij}K^{ij}$ should decay as $1/r^6$. Assuming this form, we estimate that the exterior contribution from the $K_{ij}K^{ij}$ term is 10^{-5} $M_\odot$ and can therefore usually be neglected for quasi-static neutron star orbits.

Regarding the solution for the shift vector (Eqs. (7.28) and (7.29)), we note that $\nabla \cdot \beta$ is small and changes sign across the grid. This means that the variable χ goes asymptotically to zero. Hence, we impose $\chi = 0$ along the boundary for Eq. (7.28). A solution for B^i requires that we specify the boundary condition for the "drag" component G^i. For this we note that G^i behaves as an angular momentum density and should scale along the boundary as

$$G^x = -\frac{4yJ}{r^3}, \qquad G^y = \frac{4xJ}{r^3}. \tag{7.68}$$

7.5 Orbit calculations

It is a nontrivial endeavor to find initial configurations for the two neutron stars prior to coalescence. Our method consists of placing two neutron stars on the grid with a rotational velocity sufficient to keep them in orbit and an initial "guess" density profile from a solution to the Tolman–Oppenheimer–Volkoff like equation for two single neutron stars in isotropic coordinates. The conversion from single star solution to a binary solution is achieved by allowing the stars to relax to an equilibrium configuration on the grid. That is, the field equations are solved and the hydrodynamics evolved (without the radiation reaction potential and with viscous damping of the fluid motion) until equilibrium is achieved. For the examples to be presented below, we follow the time evolution of the system with constant angular momentum until it has settled down. As the stars settle down the damping is slowly removed.

7.6 Results

The method described in this chapter is best suited for describing the orbits of neutron stars near, but not during, their final orbits. As examples, we highlight here several illustrative calculations made at selected values of the orbital angular momentum with no radiation damping of the orbits. More details of these calculations can be found in [40, 75, 76]. ·

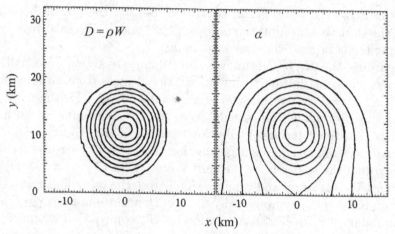

Fig. 7.2. Contours of constant density and lapse function α for an equal mass neutron star binary.

In these calculations the neutron stars are taken to be of equal mass. The baryonic mass was selected so that in isolation each star has a gravitational mass of 1.39 and 1.44 $M_\odot$. Although the calculations presented here ignore radiation damping, during most of the evolution the radiation damping is small. Therefore, the stars should follow a sequence of quasi-equilibrium configurations which closely match the equilibria computed here. These equilibria can be analyzed to obtain the rate of energy and momentum loss. Ultimately, the implied orbit decay could be used to infer the approximate time evolution through this sequence of quasi-equilibrium orbits. Figure 7.2 illustrates the density profile and metric coefficient α for 1.4 $M_\odot$ equal mass stars in a binary orbit.

In practice, the stars are placed at various separation distances on the grid and the calculation run long enough to find the stable orbit in the absence of gravitational radiation. Stars are typically followed through more than 20 revolutions to insure that the orbits have had time to settle down. We typically utilized a grid of $100 \times 25 \times 25$ zones for the matter and $100 \times 50 \times 50$ for the field variables. We make use of reflection symmetry in the orbital plane. Also, since here we study equal mass binaries, we exploit reflection inversion symmetry through the axis joining the centers of the two stars. In effect, then, this calculation is equivalent to a three-space grid of 10^6 zones.

Consider first the somewhat soft equation of state with a critical mass of $m_c = 1.575$ and stars with a baryon mass of 1.548 $M_\odot$ corresponding to $M_G = 1.39\ M_\odot$ and $\rho_c = 1.34 \times 10^{15}$ g cm^{-3} in isolation. These parameters roughly correspond to one of the simulations in Table 2 of [39]. Results

Table 7.1 Summary of orbit parameters for two equal mass neutron stars each with $m_B = 1.548$ M$_\odot$ ($m_G = 1.39$ M$_\odot$ in isolation); calculations employ a realistic equation of state [40]

J (10^{11} cm^2)	ω (rad s^{-1})	d_p (km)	ρ_c (10^{15} g cm^{-3})	U^2
∞	0.0	∞	1.34	0.0
3.0	1025	72.0	1.37	0.0080
2.8	1190	65.6	1.375	0.0085
2.6	1460	56.2	1.38	0.0097
2.4	1660	51.4	1.40	0.0115
2.2	2110	43.8	1.43	0.0145
2.1	2425	42.0	1.44	0.0165
2.0	2530	41.2	1.45	0.0185
1.9	2750	39.0	1.455	0.020
1.8	3000	35.0	1.46	0.023
1.7	3450	34.0	1.47	0.0245
1.65	4200	29.2	1.50	0.0335
1.6	(inspiral)			

Table 7.2 Summary of orbit parameters for two equal mass neutron stars each with $m_B = 1.61$ M$_\odot$ ($m_G = 1.44$ M$_\odot$ in isolation); calculations employ a realistic equation of state [40]

J (10^{11} cm^2)	ω (rad s^{-1})	d_p (km)	ρ_c (10^{15} g cm^{-3})	U^2
∞	0.0	∞	1.38	0.0
3.0	1710	56.4	1.51	0.013
2.6	1800	52.0	1.55	0.015
2.4	2100	47.2	1.57	0.018
2.2	2400	44.2	1.60	0.020
2.0	3000	37.2	1.64	0.030
1.9	3600	29.8	1.72	0.040
1.85 (collapsing)	4500	23.6	4.05	0.070

of this calculation are summarized in Table 7.1 from [40]. These stars are stable until orbit inspiral.

We did find, however, that collapse could occur prior to inspiral if the stars were increased in mass from $m_G = 1.39$ to $m_G = 1.44$ M$_\odot$ ($m_B = 1.61$ M$_\odot$) for this simulation. The results from this run are given in Table 7.2. Collapse of the stars was observed to occur for very close orbits ($J = 1.85 \times 10^{11}$ cm^2) just before inspiral. The coordinate separation between stars was only 2.4 times the coordinate radii. At $J = 1.8 \times 10^{11}$ cm^2 the orbit still appeared stable though the stars started to collapse. In another calculation we used a softer equation of state for which $m_c = 1.54$ M$_\odot$ and $m_G = 1.40$, $m_B = 1.54$ M$_\odot$. For this case, collapse occurred with $J = 2.0 \times 10^{11}$ cm^2.

Thus, collapse prior to inspiral may still be a possibility albeit for stars close to the maximum mass of a soft equation of state and for very close orbits. Such a soft equation of state is a reasonable possibility. For example, collapse would always occur prior to inspiral for typical mass neutron stars modeled with the equation of state of Bethe and Brown [9].

Another quantity of interest is the specific angular momentum J/M_G^2 as the stars approach the final orbits. $J/M_G^2 = 1.03$ for the last computed stable orbit of the sequence in Table 7.1. For $J/M_G^2 = 0.99$ the orbit is unstable. Thus, we expect that when the stars begin to inspiral, the specific angular momentum is very near unity, $J/M_G^2 \approx 1.00$, and will become ≤ 1 as the orbit plunges. Hence, the stars can immediately spiral inward to form a Kerr black hole near maximum rotation. This has important implications for the emergent gravity wave signal from the subsequent ringing.

7.6.1 Analysis

One would like to have at least a qualitative understanding of why stars in a binary system increase their central density. After all, one expects from the equivalence principle that the neutron stars should be in free fall, and therefore only subject to tidal forces which should stabilize the stars [14, 25, 32, 70, 77]. Here we present a heuristic explanation of the observed increase in density as the stars approach each other. We trace this increase to the effects of the velocity factor $U^2 = W^2 - 1$. This factor accounts for the specific kinetic energy of the orbital motion of the stars. Its effect is only manifest when a global uniform four-velocity across the star cannot be identified, and when the stars are accelerating with respect to the gradient of a background metric. It is not a simple tidal force. It depends upon the motion of the stars. Its effect is to increase the effective source strength.

From Eq. (7.10) the Hamiltonian density ρ_H has a term $U^2(\rho + \rho \epsilon \Gamma)$ which enters into the source term ρ_1 for the conformal factor ϕ. Similarly, the source for the Poisson equation for $(\alpha \phi)$ (cf. Eq. (7.22)) has a term $3U^2(\rho + \rho \epsilon \Gamma)$. Thus, the source terms for both ϕ and α will increase as the separation distance decreases and U^2 increases. A stronger source term will imply larger values for ϕ and smaller values for α at the centers of the stars, and therefore larger gravitational force terms $\partial\phi/x^i$ and $\partial\alpha/x^i$ between the centers and surfaces of the stars.

In isotropic coordinates, the general relativistic condition of hydrostatic equilibrium for each star can be inferred from the dominant terms in the momentum equation (Eq. (7.42))

$$\frac{\partial P}{\partial x^i} \approx -(\rho + \rho\epsilon\Gamma)\left(\frac{\partial \log \alpha}{\partial x^i} + \left[\frac{\partial \log \alpha}{\partial x^i} - 2\frac{\partial \log \phi}{\partial x^i}\right]U^2\right), \qquad (7.69)$$

where we have ignored the centrifugal term, $S_j(\partial\beta^j/\partial x^i)$. From this we see that the effective gravitational force (right-hand side of Eq. (7.69)) increases both because U^2 is nonzero and because the gradients of α and ϕ are more steep as U^2 increases. Note that in a coordinate system in which a killing vector can be imposed such that V^i vanishes over the star (e.g. rigid corotation [8, 36] or uniform motion [41]), this extra term vanishes. In the case of rigid irrotation, however, there appears to be a slight residual effect [12, 13, 37, 71] as it does in the case of a free fall head-on collision (cf. Chapter 6). Also note that a further increase of binding arises from the $K^{ij}K_{ij}$ terms in the field sources, but these terms are much smaller than the U^2 contributions.

From the equations for α and ϕ one can see that $\phi^2 \approx 1/\alpha$, and Eq. (7.69) can be rewritten,

$$\frac{\partial P}{\partial x^i} \approx -(\rho + \rho\epsilon\Gamma)\left[\frac{\partial \log \alpha}{\partial x^i}(1 + 2U^2)\right]. \qquad (7.70)$$

Thus, the effective gravitational source tems for the ϕ equation (7.16) have additional factors of order U^2.

Consider the innermost orbit for which $U^2 = 0.07$ from Table 7.2. In this example, these factors are estimated to augment the source for ϕ by 7% and the source for α by $\sim 14\%$.

In the simulations described above, the released binding energy was deposited as increased internal energy (Fermi and thermal). Any thermal energy, however, was assumed to be efficiently radiated away so that the stars remained cold. However, it is not necessarily true that the input energy goes into thermal energy nor that it is efficiently radiated away. If, for example, this energy were not dissipated, the stars could simply oscillate about equilibrium rather than collapse. We argue, however, that it seems most likely that such oscillations would be quickly damped relative to the time scale for inspiraling. Initially, the radial changes will be quite small, and the coupling of radial motion to thermal excitation could occur, for example, via star quakes in analogy with observed pulsar glitches. As the rate of energy release becomes more rapid and the crust melts, we speculate that the coupling of radial modes with the orbital motion, nonradial fluid motion, and tidal forces will lead to a complex excitation of higher modes and shocks which could further heat the star. Also, the coupling of radial modes with the magnetic field could damp the oscillations. As the stars become hot enough, $T \sim 1$ MeV, neutrino viscosity could also serve to damp the radial motion.

As these dissipative processes come into play it seems plausible that significant thermal energy could be deposited in the stars. If the thermal energy is efficiently radiated away, then the stars will remain near zero temperature and the previous calculations are valid. However, it is also possible that the energy may not be radiated away as rapidly as it is released, in which case the damping will be converted into both increased Fermi energy and thermal energy. An upper limit to the temperature of the star would be derived by assuming no radiative cooling during the compression. In [39] we estimated the possible heating and radiation of the stars as they adjust to the changing orbit U^2 factor and tidal forces.

To estimate the amount of heating that would occur during the compression of the stars, a spherical hydrodynamics code was written [39] with the gravitational acceleration term rewritten as

$$\frac{\partial \ln \alpha}{\partial r} \rightarrow \frac{\partial \ln \alpha}{\partial r}(1 + 2U^2). \tag{7.71}$$

Calculations were made for a sequence of U^2 values and the thermal energy induced was evaluated. For the spherical calculation done with a cold equation of state, the central density was found to be close to the three-dimensional calculations for the same value of U^2. Heating was studied by imposing an energy conserving damping term in the equations of motion. This damping relaxes the stars to their new equilibrium. The damped kinetic energy was added as internal energy. By integrating the rate at which kinetic energy is damped into Fermi and thermal energy, and estimating the rate at which this energy could be subsequently radiated away, a measure of the heating of the star could be obtained.

The results of those calculations depend upon the equation of state and stellar mass. However, the amount of released gravitational energy E_{Rel} can roughly be fit by the equation

$$E_{Rel} \approx 300U^4 \times 10^{53} \text{ erg/binary}. \tag{7.72}$$

An estimate of the rate of increase of the orbital four-velocity can be obtained [39] from the gravitational radiation time-scale. Then from this relation between released binding energy and increasing four-velocity [39, 40], the thermal energy deposition rate into the stars can be deduced in approximate analytic form [39],

$$\dot{E}_{th} = \frac{(32/5)(Mf)^{5/3} f E_{th}^0}{[1 - (64/5)(Mf)^{5/3} ft]^{3/2}},$$ (7.73)

where f is the orbital angular frequency and E_{th}^0 is the total thermal energy deposited into the stars.

In [59] simulations were made of a hydrodynamic pair plasma formed by the emergent neutrinos from neutron stars heated in this way. A range of initial energy deposited in pair plasma, $E_{plasma}^0 = 10^{51}, 10^{52}, 10^{53}$ erg, was considered. At the time a typical neutron star binary system is near the last stable obit, the orbital frequency is approximately a few $\times 10^3$ s^{-1}. Hence, by Eq. (7.73), the energy deposition rate would be

$$\dot{E}_{th} \approx 10^2 \times E_{th}^0 \text{ erg s}^{-1}.$$ (7.74)

Thus, for $E_{th}^0 = 2 \times 10^{52}$ erg, $\dot{E}_{th} \approx 2 \times 10^{54}$ erg s^{-1}.

This much heating could lead to copious neutrino emission and may provide a framework in which to produce γ-ray or X-ray bursts. In [59] it was found that this mechanism might account for observed properties of short duration γ-ray bursts, especially if one takes into account [58] the increased efficiency for neutrino annihilation into pair plasma in the region between the stars, and the enhancement of neutrino annihilation due to relativistic effects [57].

7.7 Solving the Einstein equation in three dimensions

We conclude this chapter with a summary of a promising method for the ultimate solution of the full Einstein equations with relativistic hydrodynamics. This so-called BSSN method originally due to Shibata and Nakamura and then Baumgarte and Shapiro is a currently popular alternative to the straightforward integration of the ADM equations for γ_{ij} and K_{ij}. The basic idea is to evolve separately a conformal factor and the trace of the extrinsic curvature. Details of the implementation of this method can be found in [7, 63]. Although the authors of this book have not utilized this method, it seems fitting to mention it here as at least an indication of how to approach a solution to the full field equations.

In this approach, one begins by writing a conformal metric,

$$\tilde{\gamma}_{ij} = e^{-4\phi} \gamma_{ij}.$$ (7.75)

One can then choose

$$e^{4\phi} = \gamma^{1/3} \equiv \det(\gamma_{ij})^{1/3}.$$ (7.76)

With this choice, the determinant of $\tilde{\gamma}_{ij}$ is unity. One also writes the trace-free part of the extrinsic curvature K_{ij} as

$$A_{ij} = K_{ij} - \frac{1}{3}\gamma_{ij}K, \tag{7.77}$$

where in this case, $K = \gamma^{ij}K_{ij}$. One also introduces,

$$\tilde{A}_{ij} = e^{-4\phi}A_{ij}, \tag{7.78}$$

where one can raise and lower indices of $\tilde{A}_{ij}$ with the conformal metric $\tilde{\gamma}_{ij}$, i.e. $\tilde{A}^{ij} = e^{4\phi}A^{ij}$.

Now, taking the trace of the evolution equations for $\dot{\gamma}_{ij}$ and $\dot{K}_{ij}$, with respect to the physical metric γ_{ij}, one has

$$\frac{d}{dt}\phi = -\frac{1}{6}\alpha K \tag{7.79}$$

and

$$\frac{d}{dt}K = -\gamma^{ij}D_j D_i\alpha + \alpha(\tilde{A}_{ij}\tilde{A}^{ij} + \frac{1}{3}K^2) + \frac{1}{2}\alpha(\rho + S), \tag{7.80}$$

where the Hamiltonian constraint has been utilized to eliminate the Ricci scalar from this equation. The trace-free parts of the two evolution equations become

$$\frac{d}{dt}\tilde{\gamma}_{ij} = -2\alpha\tilde{A}_{ij} \tag{7.81}$$

and

$$\frac{d}{dt}\tilde{A}_{ij} = e^{-4\phi}\left(-(D_i D_j\alpha)^{TF} + \alpha(R_{ij}^{TF} - S_{ij}^{TF})\right)$$
$$+ \alpha(K\tilde{A}_{ij} - 2\tilde{A}_{il}\tilde{A}^l_j), \tag{7.82}$$

where the superscript TF denotes the trace-free part, e.g. $R_{ij}^{TF} = R_{ij} - \gamma_{ij}R/3$.

Next, we write the Ricci tensor R_{ij} as

$$R_{ij} = \tilde{R}_{ij} + R_{ij}^\phi, \tag{7.83}$$

where R_{ij}^ϕ is

$$R_{ij}^\phi = -2\tilde{D}_i\tilde{D}_j\phi - 2\tilde{\gamma}_{ij}\tilde{D}^l\tilde{D}_l\phi$$
$$+ 4(\tilde{D}_i\phi)(\tilde{D}_j\phi) - 4\tilde{\gamma}_{ij}(\tilde{D}^l\phi)(\tilde{D}_l\phi), \tag{7.84}$$

and $\tilde{D}_i$ is the derivative operator associated with $\tilde{\gamma}_{ij}$. Then, the Ricci tensor is made into an elliptic form by introducing the "conformal connection functions"

$$\tilde{\Gamma}^i \equiv \tilde{\gamma}^{jk}\tilde{\Gamma}^i_{jk} = -\tilde{\gamma}^{ij}_{,j}, \tag{7.85}$$

where the $\tilde{\Gamma}^i_{jk}$ are the connection coefficients associated with $\tilde{\gamma}_{ij}$. The Ricci tensor then becomes

$$\tilde{R}_{ij} = -\frac{1}{2}\tilde{\gamma}^{lm}\tilde{\gamma}_{ij,lm} + \tilde{\gamma}_{k(i}\partial_{j)}\tilde{\Gamma}^k + \tilde{\Gamma}^k\tilde{\Gamma}_{(ij)k} + \tilde{\gamma}^{lm}\left(2\tilde{\Gamma}^k_{l(i}\tilde{\Gamma}_{j)km} + \tilde{\Gamma}^k_{im}\tilde{\Gamma}_{klj}\right). \tag{7.86}$$

For appropriately chosen coordinates (e.g. $\beta^i = 0$), the problem is reduced in this way to a set of coupled nonlinear, inhomogeneous wave equations for the conformal metric $\tilde{\gamma}_{ij}$. In these coupled equations then, the gauge terms K and $\tilde{\Gamma}^i$, the conformal factor $\exp(\phi)$, and the matter terms M_{ij} appear as sources.

Finally, an equation for the evolution of $\tilde{\Gamma}^i$ can be obtained,

$$\frac{\partial}{\partial t}\tilde{\Gamma}^i = -\frac{\partial}{\partial x^j}\left(2\alpha\tilde{A}^{ij} - 2\tilde{\gamma}^{m(j}\beta^{i)}{}_{,m} + \frac{2}{3}\tilde{\gamma}^{ij}\beta^l{}_{,l} + \beta^l\tilde{\gamma}^{ij}_{,l}\right). \tag{7.87}$$

One also utilizes the momentum constraint to eliminate the divergence of $\tilde{A}^{ij}$. This gives

$$\frac{\partial}{\partial t}\tilde{\Gamma}^i = -2\tilde{A}^{ij}\alpha_{,j} + 2\alpha\left(\tilde{\Gamma}^i_{jk}\tilde{A}^{kj} - \frac{2}{3}\tilde{\gamma}^{ij}K_{,j} - \tilde{\gamma}^{ij}S_j + 6\tilde{A}^{ij}\phi_{,j}\right)$$
$$+ \frac{\partial}{\partial x^j}\left(\beta^l\tilde{\gamma}^{ij}_{,l} - 2\tilde{\gamma}^{m(j}\beta^{i)}{}_{,m} + \frac{2}{3}\tilde{\gamma}^{ij}\beta^l{}_{,l}\right). \tag{7.88}$$

In this way one has reduced the problem to a set of variables, ϕ, K, $\tilde{\gamma}_{ij}$, $\tilde{A}_{ij}$ and $\tilde{\Gamma}^i$ (not all of which are independent) which can be evolved for the field. These can be related to their ADM counterparts for the solution of the hydrodynamics as described in previous chapters (see [62] for an application of this method).

References

[1] A. M. Abrahams, in *Proc. Sixth Marcel Grossmann Meeting, Kyoto, 1991*, ed. H. Sato and T. Nakamura (World Scientific, Singapore, 1992), p. 345.

[2] A. Abramovici, W. E. Althouse, R. W. P. Drever, Y. Gursel, S. Kawamura, F. J. Raab, D. Shoemaker, L. Sievers, R. E. Spero and K. S. Thorne, *Science*, **256**, 325 (1992).

[3] E. Amaldi, O. Aguiar, M. Bassan, P. Bonifazi, P. Carelli, M. G. Castellano, G. Cavallari, E. Coccia, C. Cosmelli, W. M. Fairbank, S. Frasca, V. Foglietti, R. Habel, W. O. Hamilton, J. Henderson, W. Johnson, K. R. Lane, A. G. Mann, M. S. McAshan, P. F. Michelson, I. Modena, G. V. Pallottino, G. Pizzela, J. C. Price, R. Rapagnani, F. Ricci, N. Solomonson, T. R. Stevenson, R. C. Taber and B. X. Xu, *Astron. Astrophys.*, **216**, 325 (1989).

[4] S. B. Anderson, P. W. Gorham, S. R. Kulkarni, T. A. Prince and A. Wolszczan, *Nature*, **346**, 42 (1990).

[5] P. Anninos, D. Hobill, E. Seidel and L. Smarr, *Phys. Rev. Lett.*, **71**, 2851 (1993).

[6] R. Arnowitt, S., Deser and C. W. Misner, in *Gravitation*, ed. L. Witten (Wiley, New York, 1962), p. 227.

[7] T. W. Baumgarte and S. L. Shapiro, *Phys. Rev.*, **D59**, 024002 (1999); *Phys. Rep.*, **376**, 41 (2003).

[8] T. W. Baumgarte, G. B. Cook, M. A. Scheel, S. L. Shapiro and S. A. Teukolsky, *Phys. Rev. Lett.*, **79**, 1182 (1997) gr-qc/9704024; also gr-qc/9705023; gr-qc/9709026.

[9] H. A. Bethe and G. E. Brown, *Astrophys. J.*, **445**, L129 (1995).

[10] R. M. Bionata, et al., *Phys. Rev. Lett.*, **58**, 1494 (1987).

[11] L. Blanchet, T. Damour, B. R. Iyer, C. M. Will and A. G. Wiseman, *Phys. Rev. Lett.*, **74**, 351 (1995).

[12] S. Bonazzola, E. Gourgoulhon and J.-A. Marck, *Phys. Rev.*, **D56**, 7740 (1997); S. Bonazzola, E. Gourgoulhon and J.-A. Marck, in *Proc. 19th Texas Symposium on Relativistic Astrophysics and Cosmology, Paris, December 14–18, 1998*, ed. R. Aubourg, T. Montmerle, J. Paul and P. Peter (CRA/DAPNIA, Gif sur Yvette, 2000); *Phys. Rev. Lett.*, **82**, 892 (1999); E. Gourgoulhon, P. Grandclement, K. Taniguchi, J.-A. Marck and S. Bonazzola, *Phys. Rev.*, **D63**, 064029 (2001).

[13] S. Bonazzola, E. Gourgoulhon and J.-A. Marck, *Phys. Rev.*, **D58**, 104020 (1998).

[14] P. Brady and S. Hughes, *Phys. Rev. Lett.*, **79**, 1186 (1997).

[15] B. Caron, A. Dominjon, F. Marion, L. Massonnet, R. Morand, B. Mours, M. Yvert, D. Babusci, H. Fang, G. Giordano, G. Matone, L. Matone, V. Sannibale, J.-M. Mackowski, M. Napolitano, L. Pinard, C. Boccara, Ph. Gleizes, V. Loriette, J.-P. Roger, F. Barone, E. Calloni, L. di Fiore, A. Grado, M. Longo, L. Milango, G. Russo, S. Solimeno, F. Bondu, A. Brillet, V. Brisson, F. Cleva, M. Davier, H. Heitmann, P. Hello, J. M. Innocent, M. Jacquemet, L. Latrach, F. Le Diberder, C. N. Man, A. Marraud, G. M. Nguyen, M. Pham-Tu, J.-Y. Vinet, G. Cagnoli, L. Gammaitoni, F. Marchesoni, M. Punturo, S. Braccini, C. Bradaschia, R. del Fabbro,

A. di Virgilio, I. Ferrante, F. Fidecaro, R. Flaminio, A. Giassi, A. Giazotto, G. Gorini, L. E. Holloway, C. X. Hong, A. Lusiani, M. Morganti, F. Palla, D. Passuello, R. Poggiani, G. Torelli and Z. Zhou, *Nucl. Instrum. Methods, Phys. Res. A*, **360**, 258 (1995).

[16] J. P. A. Clark and D. M. Erdley, *Astrophys. J.*, **215**, 311 (1977).

[17] J. P. A. Clark E. P. J. van den Heuvel and W. Sutantyo, *Astron. Astrophys.*, **72**, 120 (1979).

[18] G. B. Cook, S. L. Shapiro and S. A. Teukolsky, *Phys. Rev.*, **D53**, 5533 (1996).

[19] C. Culter, T. A. Apostolatos, L. Bildsten, L. S. Finn, E. E. Flanagan, D. Kennefick, D. M. Markovic, A. Ori, E. Poisson and G. J. Sussman, *Phys Rev. Lett.*, **70**, 2984 (1993).

[20] T. Damour, R. S. Hanni, R. Ruffini and J. R. Wilson, *Phys. Rev.*, **D15**, 1518 (1978).

[21] M. B. Davies, W. Benz, T. Piran and F.-K. Thielemann, *Astrophys. J.*, **431**, 742 (1994).

[22] L. P. Eisenhart, *Riemannian Geometry* (Princeton University Press, Princeton, NJ, 1966).

[23] C. R. Evans, Ph.D. Thesis, University of Texas (1985).

[24] L. S. Finn, *Phys. Rev. Lett.*, **73**, 1878 (1994).

[25] E. Flanagan, *Phys. Rev.*, **D58**, 124030 (1998).

[26] E. Flanagan, *Phys. Rev. Lett.*, **82**, 1354 (1999).

[27] K. Hirata, *Phys. Rev. Lett.*, **58**, 1490 (1987).

[28] R. A. Hulse and J. H. Taylor, *Astrophys. J.*, **195**, L51 (1975).

[29] H. T. Janka and M. Ruffert, *Astron. Astrophys.*, **307**, L33 (1996).

[30] L. E. Kidder, C. M. Will and A. G. Wiseman, *Phys. Rev.*, **D47**, 3281 (1993).

[31] D. Kramer, H. Stephani, E. Herlt and M. MacCallum, in *Exact Solutions of Einstein's Field Equations*, ed. E. Schmutzer (Cambridge University Press, Cambridge, 1980).

[32] D. Lai, *Phys. Rev. Lett.*, **76**, 4878 (1996).

[33] D. Lai and A. G. Wiseman, *Phys. Rev.*, **D54**, 3958 (1996).

[34] C. W. Lincoln and C. M. Will, *Phys. Rev.*, **D42**, 1123 (1990).

[35] J. C. Lombardi, F. A. Rasio and S. Shapiro, *Phys. Rev.*, **D56**, 3416 (1997).

[36] P. Marronetti, G. J. Mathews and J. R. Wilson, *Phys. Rev.*, **D58**, 107503 (1998).

[37] P. Marronetti, G. J. Mathews and J. R. Wilson, *Phys. Rev.*, **D60**, 087301 (1999).

[38] G. J. Mathews and J. R. Wilson, in *Proc. 3rd Huntsville Gamma Ray Burst Symposium*, ed. C. Kouveliotou, M. F. Biggs and G. Fishman (AIP, New York, 1996), pp. 768–771.

[39] G. J. Mathews and J. R. Wilson, *Astrophys. J.*, **482**, 929 (1997).

[40] G. J. Mathews and J. R. Wilson, *Phys. Rev.*, **61**, 127304 (2000).

[41] G. J. Mathews, P. Marronetti and J. R. Wilson, *Phys. Rev.*, **D58**, 043003 (1998).

[42] B. S. Meyer, *Astrophys. J.*, **343**, 254 (1989).

[43] C. W. Misner, K. S. Thorne and J. A. Wheeler, *Gravitation* (Freeman, San Francisco, CA, 1973).

[44] T. Nakamura and K. Oohara, *Prog. Theor. Phys.*, **82**, 1066 (1989).

[45] R. Narayan, T. Piran and A. Shemi, *Astrophys. J.*, **379**, L17 (1991).

[46] R. Oechstin, S. Rosswog and F.-K. Thielemann, *Phys. Rev.*, **D65**, 103005 (2002).

[47] W. Ogawaguchi and Y. Kojima, *Prog. Theor. Phys.*, **96**, 901 (1996).

[48] K. Oohara and T. Nakamura, *Prog. Theor. Phys.*, **82**, 535 (1989).

[49] K. Oohara and T. Nakamura, *Prog. Theor. Phys.*, **83**, 906 (1990).

[50] B. Paczyński, *Astrophys. J. Lett.*, **308**, L43 (1986).

[51] B. Paczyński, *Astrophys. J.*, **363**, 218 (1990).

[52] E. S. Phinney, *Astrophys. J.*, **380**, L17 (1991).

[53] T. Piran, *Astrophys. J. Lett.*, **389**, L45 (1992).

[54] F. A. Rasio and S. L. Shapiro, *Astrophys. J.* **401**, 226 (1992); *Classical Quantum Grav.*, **16**, R1 (1999).

[55] M. Ruffert, H. T. Janka and G. Schafer, *Astrophys. Space Sci.*, **231**, 423 (1995).

[56] R. Ruffini and J. R. Wilson, *Phys. Rev.*, **D12**, 2959 (1975).

[57] J. A. Salmonson and J. R. Wilson, *Astrophys. J.*, **517**, 859 (1999).

[58] J. Salmonson and J. R. Wilson, *Astrophys. J.*, **569**, 682 (2002).

[59] J. Salmonson, J. R. Wilson and G. J. Mathews, *Astrophys. J.*, **433**, 471 (2001).

[60] B. F. Schutz, *Nature*, **323**, 310 (1986).

[61] B. F. Schutz, *Classical Quantum Grav.*, **6**, 1761 (1989).

[62] M. Shibata, *Phys. Rev.*, **D60**, 104052 (1999) gr-qc/9908027; M. Shibata and K. Uryu, *Phys. Rev.*, **D61**, 064001 (2000) gr-qc/9911058.

[63] M. Shibata and T. Nakamura, *Phys. Rev.*, **D52**, 5428 (1995).

[64] G. H. Stokes, J. H. Taylor and R. J. Dewey, *Astrophys. J.*, **294**, L21 (1985).

[65] E. Symbalisty and D. N. Schramm, *Astrophys. Lett.*, **22**, 143 (1982).

[66] K. Taniguchi and T. Nakamura, *Prog. Theor. Phys.*, **96**, 693 (1996).

[67] S. A. Teukolsky, *Astrophys. J.*, **504**, 442 (1998).

[68] K. S. Thorne, *Rev. Mod. Phys.*, **52**, 299 (1980).

[69] K. Thorne, in *300 Years of Gravitation*, ed. S. Hawking and W. Israel, (Cambridge University Press, Cambridge, 1987), p. 378.

[70] K. Thorne, *Phys. Rev.*, **D58**, 124031 (1998).

[71] K. Uryu and Y. Eriguchi, *Phys. Rev.*, **D61**, 124023 (2000) gr-qc/9908059; K. Uryo, Y. Eriguchi and M. Shibata, *Phys. Rev.*, **D62**, 104015 (2000) gr-qc/0007042.

[72] N. Wex and G. Schäfer, *Classical Quantum Grav.*, **10**, 2729 (1993).

[73] J. R. Wilson, in *Sources of Gravitational Radiation*, ed. L. Smarr (Cambridge University Press, Cambridge, 1979), p. 423.

[74] J. R. Wilson, *Phys. Rev.*, **D66**, 084015 (2002).

[75] J. R. Wilson and G. J. Mathews, *Phys. Rev. Lett.*, **75**, 4161 (1995).

[76] J. R. Wilson, G. J. Mathews and P. Marronetti, *Phys. Rev.*, **D54**, 1317 (1996).

[77] A. G. Wiseman, *Phys. Rev. Lett.*, **79**, 1189 (1997).

[78] A. Wolszczan, *Nature*, **350**, 688 (1991).

[79] S. E. Woosley, J. R. Wilson, G. J. Mathews, R. D. Hoffman and B. S. Meyer, *Astrophys. J.*, **433**, 229 (1994).

[80] J. W. York, Jr., *J. Math. Phys.*, **14**, 456 (1973).

[81] J. W. York, Jr., in *Sources of Gravitational Radiation*, ed. L. Smarr (Cambridge University Press, Cambridge, 1979), p. 83.

[82] X. Zhuge, J. M. Centrella and S. L. W. McMillan, *Phys. Rev.*, **D50**, 6247 (1994).

Index